Pr
UNMA   AI

"*Unmasking AI* is an urgent call that changed me in ways I did not expect. This is a book about what it means to be a whole human and the cost of knowingly or unknowingly turning our wholeness over to technology. I have a new critical awareness about the potential of AI to scale the worst parts of us—including racial bias—*and* I also learned about why the best parts of us—poetry, justice, love, and creativity—matter more than ever. Dr. Joy Buolamwini is a poet of code, a computer scientist, and an artist, and that's exactly who we need shaping our future. AI is not coming, it's here. If we answer the beautiful call inside these pages, we can decide who we are going to be and how we're going to use technology in service of what it means to be fully human."

—BRENÉ BROWN, #1 *New York Times*
bestselling author of *Dare to Lead*

"Buolamwini's book recounts her journey to become one of the nation's preeminent scholars and critics of artificial intelligence . . . and offers readers a compelling, digestible guide to some of the most pressing issues in the field."

—*Los Angeles Times*

"Buolamwini looks at the social implications of the technology and warns that biases in facial analysis systems could harm millions of people—especially if they reinforce existing stereotypes."

—NPR

"A triumph in a literary work about artificial intelligence . . . Readers come away with a better understanding **of AI.**"

—*Business Insider*

"Joy Buolamwini joined the president for a closed-door roundtable in June. The founder of the Algorithmic Justice League (AJL) voiced concerns about facial recognition and biometrics already being used in policing, education, and health care."

—*Time* (The 100 Most Influential People in AI 2023)

"Buolamwini seems to be everywhere all at once. Her work resonates because she is both a computer scientist, with a doctorate from the Massachusetts Institute of Technology, and an artist."
—*The Boston Globe* (Bostonians of the Year 2023)

"[A] warning about the dangers of artificial intelligence."
—*Essence*

"Legendary A.I. researcher Joy Buolamwini explores how we can be more human in a world that is increasingly dominated by technology."
—*Inc.* (The Best Books of 2023)

"Joy Buolamwini describes research on racial and gender bias in the algorithms underlying machine intelligence. Her expertise now has her advising world leaders on how to prevent AI's harms."
—*The New Scientist* (Best Non-Fiction and Popular Science Books of 2023)

"Joy Buolamwini, founder of the Algorithmic Justice League, isn't content with simply uncovering biases in facial recognition technology; she's dismantling a system that amplifies societal inequities. Through groundbreaking research like *Unmasking AI,* she exposed how algorithms disproportionately misidentify darker-skinned individuals, potentially leading to discriminatory outcomes in areas like criminal justice and employment."
—*Austin American-Statesman*

"Since this discovery, [Buolamwini] has gone on to become one of the world's most inspiring, passionate, and impactful figures in AI today at a time when her expertise [is] needed so badly."
—PRINCE HARRY, presenting the NAACP–Archewell Digital Civil Rights Award

"This is as much a memoir as it is a clarion call for change. *Unmasking AI* belongs alongside Cathy O'Neil's *Weapons of Math Destruction* and Safiya Umoja Noble's *Algorithms of Oppression* as essential warnings for our time. It's an important corrective to our unquestioning embrace of technology."
—*Booklist* (starred review)

"[A] trenchant debut . . . Buolamwini proves that she's among the sharpest critics of AI, and her list of principles for achieving 'algorithmic justice,' which includes the stipulation that 'people have a voice' in shaping the algorithms that influence their lives, charts a path forward. Urgent and incisive, this is a vital examination of AI's pitfalls."
—*Publishers Weekly*

"A timely call to action about the near and present dangers of AI systems."
—*Kirkus Reviews*

"This revelatory book exposes the myriad, deeply ingrained biases encoded into facial recognition and other 'trusted' AI systems, pushing us to confront our blind trust in the machines that are taking over our lives. In describing how she conquered her own demons along her path toward justice for all, Dr. Joy Buolamwini offers a deeply felt, stirring call to action for ethical AI—a must-read for those who want a world in which technology serves humanity."
—MARIA RESSA, Nobel Peace Prize winner, CEO and president of *Rappler*

"In a world plagued by AI harms and threats to our civil rights, Dr. Joy Buolamwini has been an essential figure in bringing irresponsible, profit-hungry tech giants to their knees. *Unmasking AI* is an inspiring story about her journey from precocious student to groundbreaking pioneer in a field where women like herself are often rendered invisible. If you're going to read only one book about AI, this should be it."
—DARREN WALKER, president of the Ford Foundation

"Joy Buolamwini is a unique and powerful intellectual force, and this book explains why. We are honored to follow her transformation from earnest and diligent student to outspoken and celebrated role model for algorithmic justice, rooting for her and likewise for our combined future."
—CATHY O'NEIL, author of *Weapons of Math Destruction*

"*Unmasking AI* shows Dr. Joy Buolamwini's unmatched ability to break down complex topics for a wide audience. While taking us through her journey from a curious scientist excited by the possibilities of AI to one who uncovers its harms, Buolamwini breaks down the past, present, and future of AI harms in a manner that allows everyone to understand and participate in resisting them. This book is yet another artifact of her excellence."

—TIMNIT GEBRU, founder of Distributed AI Research
Institute and co-founder of Black in AI

"Through stories that are both personal and deeply relevant for all of humanity, Dr. Joy Buolamwini brings wit and clarity to the punishing reality of AI bias. *Unmasking AI* illuminates achievable paths for the world's future that are far more promising and just than our current trajectories."

—MEGAN SMITH, former chief technology officer
of the United States, member of the National
Academy of Engineering, and CEO of shift7

UNMASKING AI

UNMASKING
AI

My Mission to Protect What Is
Human in a World of Machines

DR. JOY BUOLAMWINI

RANDOM HOUSE

New York

2024 Random House Trade Paperback Edition

Published in the United States by Random House, an imprint and division of Penguin Random House LLC, New York.

RANDOM HOUSE and the HOUSE colophon are registered trademarks of Penguin Random House LLC.

Originally published in hardcover in the United States by Random House, an imprint and division of Penguin Random House LLC, in 2023.

LIBRARY OF CONGRESS CATALOGING-IN-PUBLICATION DATA
Names: Buolamwini, Joy, author.
Title: Unmasking AI : my mission to protect what is human in a world of machines / Joy Buolamwini.
Description: New York : Random House, [2023]
Identifiers: LCCN 2023022181 (print) | LCCN 2023022182 (ebook) | ISBN 9780593241844 (paperback) | ISBN 9780593241851 (ebook)
Subjects: LCSH: Artificial intelligence—Moral and ethical aspects. | Artificial intelligence—Social aspects. | Artificial intelligence—Philosophy. | Discrimination in science. | Sex discrimination in science. | Buolamwini, Joy.
Classification: LCC Q334.7 .B86 2023 (print) | LCC Q334.7 (ebook) | DDC 006.3—dc23/eng/20231010
LC record available at https://lccn.loc.gov/2023022181
LC ebook record available at https://lccn.loc.gov/2023022182

Printed in the United States of America on acid-free paper

randomhousebooks.com

1st Printing

Book design by Debbie Glasserman

For my family, blood and chosen, who encircle me with love

For daughters of diasporas who yearn to be seen

For the Algorithmic Justice League past, present, and future

For the excoded everywhere who soon will be free

CONTENTS

INTRODUCTION

CAMBRIDGE, MASSACHUSETTS, 2015

Halloween was approaching and so was the end of the semester. With project deadlines hovering over me, it was
looking like another late night at my laptop, when my
friend Cindy managed to convince me to join her and our friends
for a night off. I hesitated—after hours of work, I felt like I was
finally close to a breakthrough—but ultimately relented. Despite
Cambridge's enticing fall weather, I'd spent most of my time that
semester indoors, working on the final project for "Science Fabrication," one of my first-year graduate courses at the MIT Media
Lab.

The class description grabbed me right from the start: Read
science fiction and let the literature inspire you to create something entirely new, something you've always wanted to exist, even
if it seemed impractical. Just make sure you can build it in six

weeks. Classes like this were exactly what I loved most about the Media Lab—also known as the "Future Factory." I saw it as a place of escape, a cocoon, for dreamers like me to slip into fantasy and just build cool technology. The real world and its messiness felt far away and, as a young graduate student, I embraced that cocoon.

For this class project, beyond the science fiction we read that semester, I had other sources of inspiration that were closer to home. I'd always wanted to shape-shift my body like Ananse the spider, the clever trickster who appeared in stories my Ghanaian father and mother told me while I was growing up. But how could I quickly change my body into any shape I desired without making major breakthroughs in physics? Instead of changing my physical form, I decided I would try to change the reflection of it in a mirror.

A few hours before our ladies' night out, I was seated at my desk, hacking together a prototype. With a mirror-like material called half-silvered glass placed over my laptop screen, I tapped on my well-worn keyboard, experimenting with different images projected onto a black background. I pulled up an image of Serena Williams, my favorite athlete. When I saw her eyes line up with mine in the mirror, it felt like wizardry. Serena's lips and nose became mine. It was spellbinding. But it was science, not magic.

The trick worked because of the fascinating properties of the half-silvered glass. If the glass was placed on a black background— say, the black background of my laptop screen—it appeared to be a regular mirror surface. But if there was light behind the glass, the light would shine through. So, when I put an image on a black background, it appeared through the glass, while the rest of the glass remained mirror-like. (Think of it like a video filter, but instead of seeing the effect on a digital photo of you, the effect appears on the mirror reflection of your face.) After some experimentation, I had a proof of concept—evidence that my project was feasible—for what I called the Aspire Mirror.

Like most engineers I know, once I overcame this first technical challenge, I immediately dove into the next one: The mirror worked so long as I remained perfectly still, but to heighten the illusion, I wanted to get the image to follow my face when I moved.

I had been lost in my work for hours, energized by the progress I'd made, when I suddenly realized I was running late for our night out. Phase two of the Aspire Mirror would have to wait.

Apparently, a "night out" designed by MIT women was broken down into phases. The first phase was snacks and beautification. The second phase was partying in downtown Boston. As I got ready to rush over to Cindy's apartment, I tried to recall if the invite had asked guests to bring anything. I remembered the call to bring party clothes, and there was also something about masks. That made sense, I thought: It was Halloween, after all. I settled on my outfit for the night: a hot pink blazer, black dress pants, and a white costume mask I bought for the occasion.

When I got to her apartment, Cindy came to the door and gave me a warm hug.

"So glad you made it! Why are you carrying a mask?"

"I thought the invitation told us to bring Halloween masks?"

She broke out into a grin. "I meant beauty masks. But don't worry, I have enough for everyone. I grabbed so many from my last trip to Korea."

Chuckling at my mistake, I joined the other ladies in the makeshift relaxation space. Soft pillows, manicure sets, and ambient lighting accented my fellow revelers, who were reclining with beauty masks seeping into their faces. The masks didn't fit my facial features, but at least I was out of the lab.

The next day, rejuvenated from my night with the girls, I bounded back to my office and switched on the fluorescent lights. This was one of the best parts of being a coder, and an artist: the thrill of being in the middle of creating something delightful. It's like the anticipation of eating freshly baked bread after its aroma fills the room. I sat at my desk and started phase two of the Aspire Mirror project: adding interactivity and movement tracking.

Because I wanted the digital filter to follow my face, I needed to set up a webcam and face tracking software, so that the mirror could "see" me. The webcam was easy. The face tracking software was a struggle. Like many coders, I do not build everything from scratch—I rely on preexisting code, called software libraries, to create new systems. Think of it like a home improvement project. If I want to build a fence, I don't need to personally chop down trees for my posts. I can go to the hardware store and buy prefabricated items, like precut planks of wood that fit my vision. Software libraries are lines of code written by other coders, like prefabricated building blocks, and they can be downloaded online by almost anyone.

For my Aspire Mirror, I tracked down an open-source face tracking library for the project and integrated it into my code. But even when I was looking straight into the camera, the system could not detect my face. That's OK, I thought to myself. Failure was part of the process. The next question was, Could the system detect *any* face? I tested this by drawing on the palm of my hand two horizontal lines for eyes, an *L* for a nose, and a wide *U* for a smile. I held my hand in front of the camera. The software detected my elementary markings as a face!

At this point anything was up for grabs. I looked around my

office and saw the white mask that I'd brought to Cindy's the previous night. As I held it over my face, a box appeared on the laptop screen. The box signaled that my masked face was detected. I took the mask off, and as my dark-skinned human face came into view, the detection box disappeared. The software did not "see" me. A bit unsettled, I put the mask back over my face to finish testing the code.

Coding in whiteface was the last thing I expected to do when I came to MIT, but—for better or for worse—I had encountered what I now call the "coded gaze." You may have heard of the male gaze, a concept developed by media scholars to describe how, in a patriarchal society, art, media, and other forms of representation are created with a male viewer in mind. The male gaze decides which subjects are desirable and worthy of attention, and it determines how they are to be judged. You may also be familiar with the white gaze, which similarly privileges the representation and stories of white Europeans and their descendants. Inspired by these terms, the coded gaze describes the ways in which the priorities, preferences, and prejudices of those who have the power to shape technology can propagate harm, such as discrimination and erasure. We can encode prejudice into technology even if it is not intentional.

So what if my class project didn't work on me? My light-skinned classmates seemed to enjoy using it. And of course, there could certainly be an advantage to not having one's face detected, considering the consequences of cameras tracking individuals, and the dangers of mass surveillance. Though dodging the surveillance state from time to time could be an asset, what concerned me was thinking about other mistakes computers can make and who could be harmed. I felt this way because the white mask scenario is just one example of AI failure—this book will detail many others. In fact, the coded gaze extends beyond race and gender. The deeper into my research I got, the more I understood how profound and sweeping the coded gaze's impact is. It encompasses myriad ways technology can manifest harmful discrimination that expands beyond racism and sexism, including ableism, ageism, colorism, and more.

Even at the time, the kinds of AI techniques used in the code I downloaded weren't limited to class projects and seemingly innocuous labs. These systems were already jumping out of the lab and into devices and products used by, or potentially impacting, billions of people. It was 2015, and enthusiasm for AI was only growing, from consumer goods to military operations. Amazon was a year into shipping Echo, an AI-powered voice assistant that allowed users to check the weather in exchange for having a digital pair of ears listening in on conversations for the trigger word *Alexa*. In 2011, IBM had developed an AI system called Watson that beat human contestants in the trivia knowledge game show *Jeopardy!* They built on this success to develop IBM Watson Health, which launched in April 2015 with the promise to use AI solutions to "revolutionize healthcare." IBM invested $4 billion in business acquisitions to fuel the company's healthcare ambitions.* AI was being used to develop lethal autonomous weapon systems (LAWS), also referred to as killer robots, which supporters argued could be used to transform warfare. Proponents argued that more precision targeting would make military operations more efficient, thus saving resources and lives. LAWS would allow countries who could develop them to devastate other nations while suffering fewer casualties. Ethicists warned they would up-

---

* By 2022, the effort had not lived up to its promise, and IBM sold off health assets related to the Watson project. www.fiercehealthcare.com/tech/ibm-sells-watson-health-assets-to-investment-firm-francisco-partners.

end the human costs of wars in a way that could encourage more aggression. Over one thousand AI researchers signed an open letter to the UN imploring a ban on the development of LAWS.[1]

Even after encountering the coded gaze in my Aspire Mirror project, I still hoped someone else would take care of the problem. I needed to finish school, and I enjoyed the privilege of creating technology without thinking about consequences or social problems. More seasoned scholars like the AI researchers who signed the open letter against LAWS could raise these issues. I was finally in a place where I could explore my creative impulses and build technology without being bogged down by everyday concerns, let alone geopolitical issues well beyond my experience. At the time, I viewed tech critics as necessary nuisances. I knew they served an important role, but I felt they were distant from the struggles and joys of technological innovation.

Still, my encounter with the coded gaze rankled me. Sitting in my office late at night and coding in a white mask in order to be rendered visible by a machine, I couldn't help but think of Frantz Fanon's *Black Skin, White Masks*. The book, written almost a half century before my experience, interrogates the complexities of conforming oneself—putting on a mask to fit the norms or expectations of a dominant culture. After striving for years to gain entrance to this epicenter of innovation, MIT, I was reminded that I was still an outsider. I left my office feeling invisible.

In the years since I first encountered the coded gaze, the promise of AI has only become grander: It will overcome human limitations, AI developers tell us, and generate great wealth.

While AI research and development has been going on for decades, in the year 2023 it seemed the whole world was suddenly talking about AI with fear and fascination. In November 2022, OpenAI—a former nonprofit founded in 2015 (coincidentally, the same year I built the Aspire Mirror)—released ChatGPT into the world and garnered 100 million users in two months. It became the fastest-growing application at that time.[2] ChatGPT is an advanced chatbot that can reply to prompts and questions with convincing, human-like responses. ChatGPT is part of a collection of products known as generative AI. Based on a prompt, these systems can create new text, images, code, sounds, and more. The promise is that you don't have to be an illustrator or photographer to generate compelling images, just type in what you desire to see. You don't have to learn to program to build an application, just describe what you want the app to do, and the AI system will generate it for you. You don't have to be a record label to create a new song from your favorite artist; instead, use AI to output a soundtrack, remix lyrics, and generate the vocals for those lyrics. A student racing toward a deadline doesn't have to write an essay from scratch, or maybe at all—just type in your topic and the computer will provide you a draft. Generative AI products are only one manifestation of AI. Predictive AI systems are already used to determine who gets a mortgage, who gets hired, who gets admitted to college, and who gets medical treatment—but products like ChatGPT have brought AI to new levels of public engagement and awareness.

This is a transformative time, full of potential and promise. Yet the growing harms of AI—harms that go far beyond my own encounters with the coded gaze as a graduate student at MIT—remind us that the gap between promise and reality cannot be ignored. The economic potential of AI is enticing. Microsoft added $10 billion on top of an earlier $1 billion investment in OpenAI.[3] The economic threat is palpable, and so are the legal risks. Getty Images, a company that provides stock images, sued Stability AI for allegedly using its copyrighted photos to create their Stable Diffusion product.[4] Stable Diffusion, an AI image generator with competitors like Midjourney and DALL-E, allows users to create images based on a prompt like "the pope wearing a puffer jacket." The fake image of Pope Francis in a white puffer jacket that circulated online in 2023 signaled more than AI amusement.[5] Associating the head of the Catholic Church with a fashion trend is only a few keystrokes away from generating an image that could incite religious violence. Rivals and sowers of discord can use generative AI systems to create synthetic media depicting religious and political figures in false circumstances, fueling disinformation and weakening our trust in what we see with our own eyes. After "Heart on My Sleeve," an AI-generated song that simulated the voices of rappers Drake and The Weeknd, came out, Universal Music Group took protective actions. The company ordered takedowns of the viral song and issued warning shots in a statement that declared that training AI systems on the music of their artists violates copyright law.[6]

The ability to replicate someone else's voice allows for creativity that extends beyond music. On what would have been an otherwise uneventful spring day, Jennifer DeStefano picked up her phone and heard the pleading sobs of her daughter. "Mom, these bad men have me. Help me!" The caller demanded a ransom for the safe return of her daughter. Thankfully, Jennifer was able to confirm the location of her daughter shortly after the kidnapping hoax call.[7] The next target of this kind of AI hoax might be duped, as synthetic voices become easier to create and social media content provides pranksters and kidnappers with easy access to the necessary training data. Given the real harms of AI, how can we center the lives of everyday people, and especially those at the margins, when we consider the design and deployment of AI? Can we make room for the best of what AI has to offer while also resisting its perils?

None of us can escape the impact of the coded gaze. Instead, we must face it. You have a place in this conversation and in the decisions that impact your daily life, which are increasingly being shaped by advancing technology that sits under the wide—often opaque—umbrella of artificial intelligence. This book offers a path into urgent and growing conversations about the future of technology that need your voice, the voice of everyday people with lived experiences of what it means to be excluded—indeed, excoded—from systems not designed with you in mind. We need the voice of people like Robert Williams, who was wrongfully arrested in front of his children due to a false facial recognition match.[8] We need the voice of students, those struggling with e-proctoring software that flags them as cheaters.[9] We need the voice of migrants from Haiti and Africa who were caught in limbo when applying for asylum because the U.S. government required use of a mobile app that failed to verify their faces.[10]

We also need the voice of the unseen faces that do the ghost work, the data cleaning, the human translation that supports AI products. We need the voice of the parents whose children had intimate moments recorded by listening devices meant to provide hands-free convenience. We need the voice of young people educating their communities and stopping surveillance creeping into their schools. We need to remember a Belgian man who committed suicide after interacting with a chatbot. According to his widow, he would still be here had the chatbot not encouraged him to end his life.[11]

In a world where decisions about our lives are increasingly informed by algorithmic decision-making, we cannot have racial justice if we adopt technical tools for the criminal legal system that only further incarcerate communities of color. We cannot have gender equality if we employ AI tools that use historic hiring data that reflect sexist practices to inform future candidate selections that disadvantage women and gender minorities. We cannot say we are advocating for disability rights and create AI-powered tools that erase the existence of people who are differently abled by adopting ableist design patterns. We cannot claim to respect privacy rights and then have our school systems adopt AI-powered surveillance systems that reduce children to data to be sorted, tracked, and reprimanded for deviating from the algorithmic standard. If the AI systems we create to power key aspects of society—from education to healthcare, from employment to housing—mask discrimination and systematize harmful bias, we entrench algorithmic injustice. We swap fallible human gatekeepers for machines that are also flawed but assumed to be objective. And when machines fail, the people who often have the least resources and most limited access to power structures are those who have to experience the worst outcomes.

I challenge us to do much better as a society when it comes to the AI tools we create. Our standards need to exceed, not just match, the status quo, which serves only to perpetuate inequality. Most important, we need to be able to recognize that not building a tool or not collecting intrusive data is an option, and one that should be the first consideration. Do we need this AI system or this data in the first place, or does it allow us to direct money at inadequate technical Band-Aids without addressing much larger systemic societal issues?

AI will not solve poverty, because the conditions that lead to societies that pursue profit over people are not technical. AI will not solve discrimination, because the cultural patterns that say one group of people is better than another because of their gender, their skin color, the way they speak, their height, or their wealth are not technical. AI will not solve climate change, because the political and economic choices that exploit the earth's resources are not technical matters. As tempting as it may be, we cannot use AI to sidestep the hard work of organizing society so that where you are born, the resources of your community, and the labels placed upon you are not the primary determinants of your destiny. We cannot use AI to sidestep conversations about patriarchy, white supremacy, ableism, or who holds power and who doesn't. As Dr. Rumman Chowdhury reminds us in her work on AI accountability, the moral outsourcing of hard decisions to machines does not solve the underlying social dilemmas.

In seeking algorithmic justice, the option to say no, the option to halt a project, the option to admit to the creation of dangerous and harmful though well-intentioned tools must always be on the table.

The examples throughout this book and of my own personal experience reveal that AI reflects both the aspirations and the limitations of its makers. AI does not transcend humanity. Still, we can transform how AI is created to minimize the inhumane. We can also transform how we think about AI in the context of creating a more just society. I invite you into my journey from an eager computer scientist ready to solve the world's problems with code to an advocate for algorithmic justice concerned with how technology can encode harmful discrimination and exclusionary practices. I critique AI from a place of having been enamored with its promise, as an engineer more eager to work with machines than with people at times, as an aspiring academic turned into an accidental advocate, and also as an artist awakened to the power of the personal when addressing the seemingly technical.

I am a child of Ghana born to an artist and a scientist, and my background informs my sensibilities in how I learn about the world and share my evolving understanding. I aim to tell stories that make daughters of diasporas dream and sons of privilege pause. You may not agree with all the lessons I take from my observations of what ails AI. My experiences as a young Black woman may make you feel seen or feel foreign at first. You may be skeptical that there is much more to say about AI harms, or you may be eager to read the behind-the-scenes account of my public battles with tech giants like Amazon, and the more private battles too. You may still be wondering, what exactly is an algorithm and why didn't anyone tell me I could join the Algorithmic Justice League? Regardless of where you are positioned at the beginning of this book, I hope you come away with a deeper understanding of why each and every one of us has a role to play in reaching toward algorithmic justice. I hope when you feel despair you return to the stories of triumphs I share. I hope when you feel there is no place for creative expression in your work you revisit the poetry crafted for you in this book. I hope when you are afraid to speak up you read about the Brooklyn tenants who organized to resist a harmful AI system and are reminded of the value of your voice and experiences. I hope when you are tired of feelings and just want some answers, you remember this is not a book of easy answers, but I hope you walk away with questions that push us all to rethink, reframe, and recode the future of AI.

PART I

IDEALISTIC IMMIGRANT

*With safety in sight and security calling,
Would you turn back for the forgotten ones?*

*Would you risk your comfort or diminish your
power to reach out to those left in the shadows?*

*Would your lips testify of uncanny truths or
instead would you swallow your conscience and
cough up excuses?*

*I diverted my eyes because I was afraid that I
would see myself in the shadows.*

*And yes, she was there. Her name was Potential.
Her crime was hope.*

*She lived in a city that promised equality yet did
not know the meaning of equity.*

*Prestige and privilege masquerade as merit though
much of what is achieved is a function of what
we inherit.*

CHAPTER 1

DAUGHTER OF ART AND SCIENCE

I am the daughter of art and science. My mother, Frema the Akan, is the first artist I knew. As a child, I sat next to her as she filled canvas after canvas with powerful colors and made creative ideas reality. Art supplies littered our garage, mixed among drawing books, portfolios, artificial fruits, and flowers. My mother explored human conditions of the heart. Her work, she told me, was aimed at moving people to experience healing, to glimpse the divine, to be enraptured and swept into another place of awareness. I would observe her, deep in focus, considering the next stroke to apply to an evolving piece of art. Her experiments and works in progress were a constant presence for me. Seeing her sculpt, paint, draw, and etch out art was a delight to my senses. Her four-foot paintings towered over me, and the smells of charcoal and turpentine tantalized my nose. Our world was an open invitation for me to try my hand at creative expression. I soon had sketchbooks filled with whatever had recently caught my interest—ramps, skateboards, mustangs, animated characters, guitars, and amplifiers. My mother's voice of encouragement, a constant echo, gave me the audacity to explore my capacities and my curiosity. But artistic experiments were not the only ones that peppered my childhood.

My father, Dr. John Buolamwini, is the first scientist I knew. He worked on topics that would take time for me to learn to pronounce: medicinal chemistry, pharmaceutical sciences, and computer-aided drug discovery using neural nets. Trips to his lab were fun and full of many things not to touch, a lesson I often learned the hard way. Chalk is not for eating. Dry ice burns. Walking to his office I would see hallways filled with scientific posters as he waved to colleagues and students. And then if I was lucky, while he worked on the latest grant, research paper, or other desk work, I got to play on one of the computers. When he walked to the freezer in his lab I trailed him like a shadow. He put on purple gloves, pulled out a tray, and placed it on a lab bench. I struggled to get the oversized gloves on my hands as my dad beamed at my efforts. Once I was protected, he placed a pipette in my grasp and gently applied pressure on my right thumb with his. Liquid drops bathed the cancer cells beneath our cradled hands, while my eyes widened with fascination and his beard tickled my head. Next to the lab bench sat more computers. He would show me machines linked to concepts like *flow cytometry*. I would look at the squiggles on the computers that I would later learn to call graphs.

Like my mother, he was working on experiments that required bold curiosity to ask unexplored questions. But while my mother asked questions of colors, my father asked questions of cells. In the midst of their explorations, I began to ask questions about computers. For instance, how did the images that I saw on the scientific papers come to be? They looked like abstract paintings to me. My dad showed me the software on his huge Silicon Graphics computers that would create these images containing ringlets and rods of bright reds and blues, representing different protein structures. The goal of feeding the cells, designing medicinal drugs on the computer, running all the tests, and scrutinizing the squiggles was to help people who were struggling with different conditions, from heart disease to breast cancer. He showed me the software to introduce me to chemistry, but I found myself more and more enamored with the machines themselves. I quickly found games like Doom and Cycle that came preloaded. I listened to the whirs and beeps of a dial-up connection. In that office, I opened Netscape, my first browser experience into a portal I would later learn was the internet.

And so it was that, surrounded by art and science from a very young age, I was emboldened to explore, to ask questions, to dare to alter what seemed fixed, and also to view the artist's and the scientist's search for truth as common companions.

My parents taught me that the unknown was an invitation to learn, not a menacing dimension to avoid. Ignorance was a starting place to enter deeper realms of understanding. At some point, though, they would tire of my endless questions; after entertaining my curiosity for some time, my mother would sometimes bring me back down to earth with a gentle "*Why* has a long tail . . ." elongating her words as she spoke. In addition to my parents, I turned to another source of knowledge, television. As first-generation immigrants settling into Oxford, Mississippi, my parents wouldn't let me watch commercials: They wanted to shield me from the materialism that appeared to be the backbone of American culture. "You will never find your worth in things," they cautioned. However, they encouraged me to watch educational programs, so PBS became the television channel of choice in our home. I soon found myself anticipating shows like *Nature*, *National Geographic*, *Nova*, and *Scientific American Frontier*.

There was one episode in particular that left a lasting impact on me. When I was around nine years old, I watched a segment about robots. The program host visited a place called the Massachusetts Institute of Technology. He spoke to a graduate student named Cynthia Breazeal about her work on what she called social robots. Unlike the industrial robots I had seen before—hulking machinery set to tasks like stamping out sheet metal—her social robot was not focused on work but on connection and communication. She sat next to a robot she had built named Kismet, a dazzling and intricate web of metal and wires topped off with enchanting eyes, animated ears, and a cheeky smile. The moment I saw the machine appear to come to life, I was mesmerized. Could I make something like Kismet? Could I go to a place like MIT, the ever-present backdrop to so many of the science and technology shows I watched? From that moment, I decided I wanted to go to MIT and become a robotics engineer. I was blissfully unaware of any barriers or requirements. I had more questions to ask of computers, nurtured in the incubator of youthful possibilities by the belief that I could become anything I imagined.

My first step toward building robots was learning how to program machines to do what I wanted. To give machines instructions, I discovered different kinds of programming languages. I started by learning the basics of HTML and CSS to build a website. These programming languages focused on structure and formatting. HTML allowed me to define the elements I wanted to see on a webpage, like a block of text, a button, or an image. CSS let me determine what these elements would look like, from the color of the text to how much space existed between elements. Each programming language had its own rules for how to give a computer directions. Soon enough I was using these skills to code websites for my high school sports teams and make some pocket money or barter. Even if I was a benchwarmer on the basketball team, at least I did not have to pay for my uniform or shoes.

I wanted to go deeper than websites, and I was curious about how to make games like the ones I played with my brother on his Nintendo 64 or Tony Hawk Pro Skater 2, which I enjoyed on my Sony PlayStation. So I learned another programming language called Java. Here, I was introduced to the concept of an algorithm. An algorithm, at its most basic definition, is a sequence of instructions used to achieve a specific goal. To make my character move around the screen, I would write code that followed a logical sequence. For instance, if the user hits the left arrow, move the character left on the screen. Algorithms like this, as I would eventually learn, would become the basis for more powerful and dynamic systems.

I FOLLOWED MY DESIRE TO WORK ON ROBOTICS INTO COLLEGE. By my third year at the Georgia Institute of Technology in Atlanta, I was working on social robots. One of the professors I worked with, Andrea Thomaz, was a former student of Cynthia Breazeal's. And to my delight, when I started working on Thomaz's robot Simon, I learned that the code that was used to power it had descended from the *CREATURES* code library that once animated Kismet. My assignment with Simon was to see if I could have the robot engage in a social interaction with a human. I settled on working on a project called Peekaboo Simon. The aim was to have the robot participate in a simple turn-taking game with a human partner, similar to one that a parent might play with a young child. The larger aim behind this project was to see if we could have a robot play social games with young children and analyze how children responded and behaved during those interactions, thereby helping to diagnose early developmental delays or even early signs of autism. This kind of early detection could help a child receive necessary support as soon as possible.

To make this game work, I would need to get Simon the robot to detect a human face and direct its head toward the person. This was my introduction to face detection.

In class I was learning about the field of computer vision, which enabled machines to perceive the world through digital cameras and then use software on the camera input to do all kinds of tasks, like detecting a ball or a block or, in my case, a human face. When an object was detected, a rectangle, also known as a bounding box, would appear on the image to indicate its location. As I worked on the project, I ran into an issue: Peekaboo doesn't work if your partner doesn't see you, and Simon the robot was having a hard time detecting my face. I would turn on all the lights, tilt my head in many directions, and watch in dismay as the bounding box that was supposed to indicate the location of my face appeared only from time to time. In desperation, I recruited my roommate, a jovial southern woman with bright red hair, green eyes, and freckled skin. The software worked on her fair face, which let me continue with the project. I didn't think too much of the experience, as my main aim was to get the project done, and this was certainly not the first time cameras had failed me. Old childhood photos revealed less than flattering images when I would be photographed outside studio lights. In some of these photos you could see only the whites of my eyes and teeth and not the rest of my facial features.

Beyond computer vision research, I had other interests as I explored college. I tried my hand at entrepreneurship. In 2011, two of my classmates, Sarah and Elizabeth, and I entered an international competition and were among the teams selected to represent the United States in Hong Kong. Our project was focused on creating a platform that would allow musicians to have jam sessions with anyone anywhere at any time. While we did not advance past the first round, we did take full advantage of our week in Hong Kong. At night we danced hard to dubstep, and during the day we visited the Hong Kong Science Park with the pulsing beat of the night before still drumming in our heads. During our visit to the science park I met another descendant of Kismet. Autom was a healthcare social robot created by Corey Kidd, another former student of Cynthia Breazeal's, who had moved to Hong Kong to build a start-up around the idea of integrating social robots into healthcare. The technology was moving outside of the lab and into the real world.

Corey and his team demoed the robot, and I volunteered to try it out. I peered into Autom's camera eyes, but the machine struggled to detect my face. Corey seemed surprised. I suspected I knew the issue. We began to talk shop, and I found out that Corey was using the same face detection software that I had used for Simon. Thousands of miles from my dorm room in Atlanta, I encountered the same problem, but again I didn't think too much of it. There was still more dancing to do and new people to meet from all over the world.

After I finished my computer science degree from Georgia Tech, I traveled to Zambia as a Fulbright fellow to teach youth how to code, before heading to Oxford University, where I would turn my attention back to academia and technology. When I was finishing my time as a Rhodes Scholar at Oxford, I had a conversation with my favorite scientist. My dad, ever conscious of academic credentials, reminded me that I had not yet earned my PhD and should consider applying to graduate school. His call made me think of my family's legacy. Before my father was my mother's father, who earned his PhD in England in 1969, decades before I was born in 1990. I remembered my childhood dreams, and the allure of a robot named Kismet on the family television screen, and so, when graduate school application season came, I applied to only one place.

CHAPTER 2

THE FUTURE FACTORY

Some dreams do come true. After weeks of anxiously checking my email, I received an offer letter for a research assistantship at the MIT Media Lab from Ethan Zuckerman, the director of the Center for Civic Media. When I connected with him on the phone, he greeted me in my first language, Twi, and I was touched by his efforts to make me feel welcome.

Beyond Ethan's enthusiasm, the Media Lab for me held a mythical aura as the "Future Factory," a place where designers, scientists, and engineers came to dream and create possibilities of what life could be. This was the incubation hub of everything from social robots, LEGO Mindstorms, and visual programming languages to Guitar Hero, digital ink, and early sketches of autonomous vehicles. On a tour of the campus prior to accepting MIT's offer, I remember talking to a professor who told me, "If what you are thinking of making already exists, go elsewhere."

The idea of being at the vanguard of future technological breakthroughs excited me. When I arrived in 2015, the MIT Media Lab was made up of more than twenty smaller labs, including Opera of the Future, Lifelong Kindergarten, and Tangible Media, headed by famous faculty members like Cynthia Breazeal. Though most of the groups were focused on worlds yet to come, the group I joined, the Center for Civic Media, or Civic Media for short, was focused on the impact of technology on society today. Civic Media maintained that the heart of technology must center society. This made us a curiosity in an environment that was otherwise focused on transcending the limitations of society. Despite my desire to build futuristic technology, my work prior to coming to the Media Lab had been focused on technologies with immediate real-world application. For instance, as part of my undergraduate work at Georgia Tech, I worked in Ethiopia on mobile surveying tools to help combat neglected tropical diseases, creating software that changed paper-based collection methods into more efficient electronic methods. And Civic Media felt to me like the heart of the Media Lab, both in its mission and in where it was situated. We were centrally located on the third floor of a six-floor building, positioned between the original Media Lab building and the new jeweled construction, a bridge between the past and the present.

WHILE I WAS EXCITED TO BE AT MIT, UNDER THE CARE OF A SU-pervisor who wanted to make me feel like I belonged, I still encountered reminders that I was an outsider.

My first semester on campus coincided with the thirtieth anniversary of the lab, and celebrations were in full swing. Martha Stewart had come to town as part of the festivities, as had the magicians Penn and Teller. Their presence was fitting for a place that the leadership described as animated by magic, mystery, and mischief. As students, we received thirtieth-anniversary dress shirts from Ministry of Supply, a recent MIT spinout company that used astronaut fabrics. At the time, the company did not make clothing cut specifically for women, so I settled for the extra small men's dress shirt. In those days, the majority of the Media Lab faculty and students were men. Nonetheless, this wrinkle-free and stain-resistant white dress shirt would become a favorite staple of my Media Lab wardrobe, something I'd wear over more colorful red and yellow V-neck shirts. By now I was used to lending splashes of color to otherwise whitewashed vistas.

Like the dress shirt, the Media Lab was maintained to be stain free. I remember one day walking into the atrium and watching a man dab white paint on small scuff marks on the otherwise pristine walls. With this attention to detail, it is no surprise that the Media Lab was also an architectural jewel, looking more like the headquarters of a hotshot technology company than a typical academic building composed of classrooms and offices. Throughout the building, there were enticing displays of self-pedaling tricycles, metal origami blocks, and screens cycling through video demonstrations. One of my favorites was a room-sized food computer prototype. It resembled a greenhouse with different areas for various plants, each plant patch equipped with sensors that could alert the gardeners about what was needed. For example, you could receive a message on your mobile device that water levels were low and it was time to shower the tomatoes. The sensor-soaked greenhouse was illuminated with purple lights. Across the Charles River you could pick out the Media Lab from the rest of the Cambridge skyline at night just by locating the slanted silver building, glowing purple.

One night early in the semester, I called for an Uber to take me to campus, but after spending a short time with me in the car, the driver decided he did not want to drive me. When he threatened to call the cops if I did not get out, I left and started walking toward MIT. On my way there, a cop car pulled up behind me, rolling slowly nearby in warning. Was this just in my head, I wondered, or was I not completely welcome in Cambridge? At least I had security at the lab—or so I thought.

ONE MORNING DURING THAT FIRST SEMESTER, I WALKED INTO our group space. Our lab was an open area the size of half a basketball court, with colorful posters announcing events and student projects, and four wooden tables congregated into one massive gathering place in the center. At the center of the tables lay Ethan in lament. His hands gripped his Benjamin Franklin–styled hair, the graying strands cascading from his knuckles. I walked away slowly so he wouldn't know I had seen his anguish. Later I spoke to other group mates to learn what was going on. The rumors about our group were true, I'd realized. While many groups at the Media Lab were supported in part by member companies that paid at least $250,000 a year to have early access to the research at the Future Factory, our group was supported mainly by foundation grants. The future of the Civic Media group was precarious because of funding concerns. Not long after I saw him on the table, Ethan was kind enough to let me know, so I could think about alternative plans should the group not make it for the duration of my time at the Media Lab. He told me that it would be a good idea for me to form relationships with other faculty in case there was not enough support to keep our group alive.

Later in his office he asked me, "Why do you even want to be an academic in the first place? You have so many other options." His current situation was a case study in the precarity of pursuing an academic path. I leveled with him. "Ethan, I need all the credentials I can get. You have a bachelor's degree and were given the opportunity to be a professor and supervise graduate students. I don't have that same privilege." From the beginning I didn't mince words with Ethan, though most of the supervisor-student relationships around me modeled more deference.

While we didn't have the most security in terms of funding, we did have a strategic advantage. A prominent feature in the lives of ever-hungry graduate students was the Foodcam emails. The Media Lab often hosted catered events and, to reduce waste, the leftovers were placed on a black countertop in the designated food camera area. The higher profile the event, the better the food. The camera was linked to an oversized black push button. When the button was pressed, a photo of the food on offer was emailed out. Because Civic Media was the lab positioned closest to Foodcam, our graduate students were first in line. People tried to stay civil at the beginning, fast walking if not outright sprinting to the black countertop. The older graduate students brought Tupperware. Our group may have been underfunded, but we were well fed.

AS A FULBRIGHT FELLOW IN ZAMBIA IN 2013, I HAD WORKED with young people to build mobile applications designed to share information on women's rights. This work was focused on increasing the participation of women and people of color in computer science and software engineering. This dual interest in *who* was creating technology and applying technology to help solve real-world problems continued into my MIT research. When I perused the menu of potential courses to take, I went to Ethan for advice. "Take a class that builds your skills, take a class that deepens your knowledge, and take a class that is just for fun." With this push, I enthusiastically signed up for a fun course that would change the trajectory of my life: Science Fabrication. Taught by Dan Novy, a Hollywood special-effects guru, and Joost Bonsen, a catalyst for student entrepreneurship, the class was specifically focused on building fantastical futuristic technologies.

And so in my first year at the Future Factory, I started building the Aspire Mirror, the fanciful device inspired by stories my parents told me of Ananse the trickster. My white mask sat on my desk, an ever-present reminder of my most recent encounter with the coded gaze. As with my encounters with the Uber driver and the Cambridge cops, I pushed down my discomfort and carried on to the next semester. For another course, I partnered with a group of students on a project called Upbeat Walls. We were an eclectic group, including a musician, a business student at MIT's Sloan School of Management, an undergraduate studying computer science, a Media Lab master's student with engineering skills, and me, the resident computer programmer. We started with the provocation: What if you could paint walls with your smile? From there: What if your face became a paintbrush and each stroke of the brush had its own musical sound? We worked on a prototype over the course of a week.

By the time the demo day arrived, we had a system that used face tracking technology to allow a user to walk up to our small wall, move their face, and have digital strokes appear on the wall. Each stroke would come with a sound, including a beatboxing mix I had recorded.* During the demo, as with the Aspire Mirror, the system worked best for those who had lighter-skinned faces. The coded gaze lingered.

Even while I was trying to offer an escape from the ills and stresses of everyday life, I could not evade the recurring issues with the technology I was making. The software libraries I incorporated were not optimized for people like me with darker skin. I had encountered these issues when I was an undergraduate, face-to-face with Autom in Hong Kong, and here they were again. Despite all the developments in AI over those intervening years, the problem had still not been fixed. Even though I had applied the face tracking software to another domain, the underlying issues remained, clinging to me like a persistent mosquito.

---

* You can watch a demo of the UpBeat Walls on YouTube: www.youtube.com/watch?v=Pqc9tWQdW8U.

CHAPTER 3

BREAK THE ALABASTER

When I returned to MIT for the start of my second year, change was in the air. The 2016 presidential campaign was underway and a major new study about the police use of facial recognition technology had been released by Georgetown Law Center on Privacy and Technology. According to the study, the faces of one in two adults, or at least 117 million people in the United States, appeared in databases that could be searched by police departments without warrants using facial recognition technology that had not been audited for accuracy.[1] I could no longer tell myself that the issues I had experienced over and over again with computer vision technology did not warrant my specific attention. It was one thing if I could not paint a wall with my smile, but if someone was misidentified as a criminal suspect and falsely arrested or worse, then the stakes were unmistakably high.

At the same time, I was grappling with the intersection of privilege and oppression while living in relatively luxurious circumstances for a graduate student. During my time as a master's student at MIT, I was fortunate enough to become a resident tutor at Harvard. There, I helped students with fellowship applications and took a small cohort of sophomores under my wing. In exchange, I received housing, a meal plan, access to Harvard facilities, and transportation.

My apartment was on the corner of Bow and Arrow streets inside the oak-paneled walls of Westmorly Court. Outside the building, a gray stone engraving between two sets of brick-lined steps proudly announced that the site had been the student quarters of U.S. president Franklin Delano Roosevelt. It was a five-minute walk to the Charles River and an even shorter walk to the blue Harvard shuttle I caught each morning to MIT. In the fall of 2016, the Harvard dining hall workers were striking for better conditions and pay. From my room I could hear them chanting "No justice, no peace," buckets banging in punctuated rhythm, and I could see their signs from my window. I nodded my head to the beat.

Change was permeating within campus walls, as well. Media Lab graduate Karen Brennan, a Harvard professor teaching a course open to MIT students, challenged my classmates and me with the question: "What will you do with your privilege?" Maybe it was time for me to speak up about the issues I was seeing relating to the face-based technologies that were now being used by police. But I was still on the fence. I wanted to focus on being a maker, not a critic. The white mask episode had been disheartening, but I didn't want people to think I was making everything about race, nor being ungrateful for rare and hard-won opportunities. Speaking up had consequences.

Ethan, too, had been encouraging me to speak more directly to the political implications of the work I was doing. He challenged me to draw sharper connections between the lack of representation of women and people of color in the tech world and white supremacy and patriarchy. During group meetings he would say, "And Joy's work is looking at inequalities and power asymmetries in computer science education. Do you have more to add?" Uncomfortable with this framing, I demurred. "Umm, I am just focused on teaching people how to code apps they care about." Graduate school was hard enough. Why deal with the stigma and extra scrutiny of being a gadfly going on about power dynamics?

I had gotten into computer science in some ways to escape the messiness of the multiheaded -isms of racism, sexism, classism, and more. I was acutely aware of discrimination in my own life. I just wanted to embrace the joy of coding and build futuristic technologies, or even real-world applications that focused on health, without needing to be bothered with taking down -isms. I also did not want to become a nuisance. Though all signs indicated otherwise, I wanted to believe that technology could be apolitical. And I hoped that if I could keep viewing technology and my work as apolitical, I would not have to act or speak up in ways that could put me at risk. Plus, I thought focusing on these ever-present factors too much would make me bitter and rob me of the jovial spirit that had driven me to optimistically pursue male-dominated fields. Work hard enough, I believed, and you could transcend the -isms or at least minimize them to a point where your life trajectory wasn't significantly altered by their gravity. There must be some truth to meritocracy. Hadn't I made it to MIT? I wanted to hold on to that fiction instead of facing the harsher realities.

I remember arguing with my dad when I was in high school about the need for affirmative action. At the time, I didn't think it was necessary, since I had been relatively insulated from the ways that women and people of color were systemically precluded from opportunities. To help me grow out of my adolescent ignorance, my dad encouraged me to watch *Tony Brown's Journal* and a PBS women's roundtable. Regardless of race and gender, my qualifications stood on their own. There was no doubt in my mind that I was qualified, and I didn't want people to assume that I got opportunities only because of my race and gender. At the end of the day, it was a matter of personal pride. The more I grew, the more I could recognize the privileged educational opportunities I had at a young age, when I visited art galleries and science labs. My exposure to science and technology was the exception, and exceptionalism would not change the overall fabric of society.

Exceptionalism also carried the danger of tokenism, which allowed systemic issues to be ignored by pointing to a few examples of supposed success while ignoring the more common story. I was often a poster child of progress, appearing in college marketing materials and conference brochures to show that change was possible. Still, I knew that given another set of cards, my life trajectory could have been very different. There are many hardworking, brilliant people who could be in my place if they'd had similar opportunities. I had to concede that many factors outside my control contributed to my ability to take advantage of educational pathways. In many ways, I was born lucky. This acknowledgment doesn't take away from my hard work, but it does expand the conversation to be less about one individual and more about the societies we have created. And maybe given all the privileges, I did have some obligation to speak up. Maybe Ethan was right. I needed to find my voice.

Around this time, Natalie Rusk from the Lifelong Kindergarten group reached out to me about a reading at the Harvard Book Store. Author Cathy O'Neil was in town to talk about her latest book, *Weapons of Math Destruction*. Conveniently, I lived around the corner from the Harvard Book Store, so, after dinner in the Adams House dining hall, which now had a shortage of workers due to the ongoing protests, I walked over to the book talk. At the back of the store, perched on a stool, was a formidable woman with short blue hair. She talked about how data was being used to sort and control people. She highlighted the ways in which mathematical models were being used as smoke screens to obscure inequality. She explained this inequity using examples from the 2008 mortgage crisis and predatory ads for for-profit schools. During the question and answer session, I introduced myself and said that I was noticing some of what she was discussing in recent work I was doing with face detection models. Though she was not familiar with the computer vision space, the conversation I had with her made me feel I was not the only one obsessing over glimpses of the coded gaze. While I had a residential community at Harvard, sharp lab mates at MIT, and someone I was seeing who took me outside of the Cambridge bubble to spend time across the Charles River in Roxbury, I felt intellectually isolated by my growing interest in the harms of data-driven technology. I bought Cathy's bright yellow book christened with her signature. As I read the book chapter by chapter, I no longer felt alone in my exploration of algorithmic bias. If I was to be a gadfly, I might have an ally.

My thinking about technology and society was further enriched by civic-minded elders and leaders. In addition to our lab staff and graduate students, we would have regulars at weekly group meetings open to the public, including retirees like Saul and Anne, who would challenge some of the graduate students' idealistic project ideas.

"What happens to the data you are collecting about people's location?"

"How would someone without a smartphone be able to benefit from what you are proposing?"

We would also have visitors including activists, former political prisoners, and others who had been on the front lines in conflict areas. I remember being riveted by Esra'a Al Shafei, who came from Bahrain and spoke about the platforms she had built, like Mideast Youth, which gave space for music to flourish and helped queer youth find community. She spoke of her friends, fellow activists who had been imprisoned for standing up for their beliefs. She also spoke about the difficulties of fundraising and how foundations would direct her to learn from young Western men who were building platforms that were not relevant to the work she and her colleagues had done for many years. We would have visitors from places like the Robert Wood Johnson Foundation who wanted to know how our Media Cloud tool, developed to track media outlets from all around the world, was capturing information about specific topics like mental health.

Our eclectic group of visitors enabled me to see that the Center for Civic Media leveraged technology as a way to convene decision-makers and community members to have meaningful conversations that ultimately looked at the power dynamics that informed decision-making. Yes, a technical tool could help gather data, provide supporting evidence, and even enable us to answer questions we couldn't before, but this was aimed either to hold decision-makers accountable or to help connect people who cared about similar topics. For instance, this was the case with our Promise Tracker tool that was deployed in Brazil and had been used in many communities. The tool enabled novices to create community surveys and take photos to document whether public services were being delivered. So if an elected official claimed to have put money toward improving school lunches, students could document the quality of lunches they were receiving—or if they received any lunch at all—using the Promise Tracker tool. The technology was never the main point, but it was a starting point to help shift power and resources.

Still, I could not ignore my persistent feeling that it might be better for me to change my research direction. For my thesis, I had originally pursued the idea of sparking the civic imagination through mobile app development, and I had already put in significant work: I created an online course after raising funding, renting a studio for three days of shooting, and working with an editor for months. I held workshops with a local school during the summer after my first year at the Media Lab. But I couldn't shake the feeling that there was something else I was meant to be doing.

So I reached out to older graduate students to ask for their advice. I heard many arguments for not investigating the coded gaze: "If I were you, I would not change directions now; you just spent a year on a project and you only have one more year left." And "Working on topics like gender bias will receive a lot of blowback. You can get pigeonholed in ways that hinder your career." And "Your advisor is not focused on AI; you are going to have to learn so much on your own if you go down this path." The older graduates gave me sensible advice. However, I was still drawn to the bigger risk.

In high school, my favorite track event was pole vaulting. Not only did I enjoy the physical challenge, which required the speed of a sprinter, the strength of a thrower, and the body awareness of a gymnast, but it also gave me a metaphor for life. I learned early on with pole vaulting that where you fix your eyes is where you ultimately land. Staring straight at the bar often led to colliding into the bar or just barely making it over. To execute a beautiful vault, you had to look beyond the bar and rise above it to the sky. Switching your mindset from bar gazing to star gazing allows your body to follow a more expansive vision. The older graduate students at MIT who had offered me their advice were bar gazing in their attempt to finish. I still had time. I wanted to star gaze.

Ethan's office had a window facing a courtyard and another one that peeked into our lab space. The in-facing window allowed fellow graduate students to see if he appeared free for a talk. That day the blinds were up, and I could see that he was free, so I approached. Sitting on the black futon Ethan sometimes slept on when he stayed late at the lab, I let out a long sigh. I fidgeted with my hands and stayed silent for a while. Then, in a long-winded monologue I shared that I was increasingly being pulled by facial recognition technologies and that I sensed that the topic could have an important impact. I didn't know exactly what I would focus my research on, but I couldn't deny my growing conviction that there was some work in the space I was meant to do.

He looked at me with compassionate eyes. "You know, it's OK not to have all the answers, Joy."

I looked away so he wouldn't see my eyes watering.

"I know."

Being combative was easier than being vulnerable. After some reflection, he told me that as someone who would always have many interests, I would have to figure out what should remain side projects. He suggested that this new work could remain one of my artistic explorations so I could continue the momentum of the work I was already doing.

YES, THERE WAS A PART OF ME THAT LIKED MAKING COOL TECH-nology for the sake of it. But as I learned in my time at Oxford University and in Lusaka, Zambia, I was also drawn to work focused on addressing societal ills, and I knew how to use technology as an entry point into important conversations that otherwise were being ignored. Now, at the Media Lab, I was no longer focused on teaching novices how to code, as I had been when I was a Fulbright fellow. I was daring to recode artificial intelligence.

If I switched directions, would it mean all the work that I had already done was a waste of time and resources? I looked to Frema the Akan for guidance. I had watched my mother cut up some of her most gorgeous paintings that I had planned on inheriting, forcing me to think beyond the final product and focus instead on the process of creation. On another occasion, I came home from middle school and saw fragments of alabaster stone scattered throughout the dining room. That morning the pieces had been part of one holistic soft-pink sculpture that had become a companion in our home garden. I enjoyed running my hand over three smooth curves that seemed to hug each other. I would look at the sculpture with pride, feeling lucky to be the daughter of someone who could make such beautiful art. Now those embracing curves were interrupted. The sculpture was disfigured. I tried not to get too attached to my mom's artwork, though throughout our home were pieces I had grown up with, pieces that had witnessed birthday celebrations and joyous Christmas mornings. My mother modeled impermanence with the quiet confidence that came from knowing that she would inevitably make something better.

Perhaps I didn't have to hold on to the work I'd already done and instead could see value in what I had gained through the process of pursuing an idea. Because there was another question gnawing at me. The white mask haunted my thoughts.

CHAPTER 4

SHIELD READY

In addition to presenting us with the challenge: What will you do with your privilege?, Karen Brennan allowed her students to propose any project we desired, so long as it applied the concepts we explored in class. I hadn't forgotten my encounters with the coded gaze, and I had been puzzling over how I might start sharing the idea of the coded gaze with others. Still unused to the idea of speaking out about bias in technology, I decided to use Karen's class as a soft landing place to experiment with presenting my experience to a wider audience.

One day in the shower as I was brainstorming, I started pretending I was talking to a camera and began to ask questions with different voice styles. The first voice contained my initial feelings.

"Hi, camera. Can you see my face?" Scowl. "You can see my friend's face, but what about *my* face?" My original tone was pregnant with frustration and anger. When I heard myself, I wondered if my message would be diluted if I was perceived as someone holding a vendetta. I tried a different voice, which in my head sounded more like a curious researcher. After toweling off, I opened my laptop and started recording. In my ears, the recording, unlike my raw shower delivery, came off less like a rant and gained a bit more levity. I practiced not sounding angry. It took many recordings to erase subtle hints of resentment from my voice. Approaching the issue with palpable anger, I thought, would put people on the defensive when my aim was to figure out ways to have people listen. I was acutely aware of being stereotyped as an angry Black woman, eager to play the race card and find offense in the seemingly innocuous.

Even if my feelings were justified and my observations hinted at potentially larger problems, everything I had learned about navigating halls of privilege pushed me to be as amiable as possible. Just as I was masking my face to be seen by machines of silicon and steel, I was also masking my feelings to be heard by machines of flesh and bone. To do this work, I was going to need to learn to wear many different masks. If I became consumed with resentment, I would not be able to do the necessary work. I had to hold on to the belief that change was possible.

One day, as I was sitting in the Media Lab atrium on the third floor, a spry young man with curly hair came up to me. He introduced himself as Adam Horowitz and said that he was working on a pop-up art show for the Museum of Fine Arts (MFA) in Boston. He'd heard about some of my work and wanted to know if I might be interested in being part of a collaboration with the museum. The art show was to be held only a few days before Critique Day, also known as Crit Day, at the Media Lab—an annual rite of passage where second-year master's students declare their research focus and lay out their plans for making a contribution.

Year one was intentionally exploratory. While class projects, research projects, or summer work could seed the work you declared you would do, part of the beauty of the early days was to dabble. Year two was time to focus. In our introductory classes we were told that the quintessential master's thesis should be aesthetic enough to be in a museum and impactful enough to lead to the creation of a start-up, while also making a meaningful technical contribution via a new method of doing something or some other innovation. In our proposal we were required to not only show technical contributions but also speak to the societal impact of the work we planned. The MFA art show would give me a chance to try my hand at the aesthetic portion of the ideal Media Lab master's project.

For the art show, part of the MFA Now program, I was given two spaces. One was a dining area cleared away for the overnight event. The other was the main entrance wall. The only stipulation was that nothing could be attached to any walls: In true pop-up fashion, whatever was put up needed to be taken down immediately afterward, so the MFA could return to its unadorned style. I assembled an art crew of friends. Ethan put aside a budget that allowed me to purchase materials for the show to create even bigger Aspire Mirrors than the one I had made for the science fabrication class. I also worked with Rossi Films to make a mini documentary called *The Coded Gaze: Unmasking Algorithmic Bias*. The short film needed footage, and on my own I did not have the required camera skills, but I was in the ideal place to find those who could make a camera sing and stay in focus.

By chance, while walking through the third-floor atrium of the Media Lab, I met two people from the Collaboratorium, an organization focused on creative collaborations. I shared with them my idea to create a short video about the concept of the coded gaze. They were in town for a few days and agreed to help film scenes for the mini documentary for free. I went home to grab a bright red shirt. Given that this was a special occasion, I even put on a little bit of makeup, a rarity in my general rush-out-the-door routine. Late at night we met at the center Civic table, cleared away chairs, and jury-rigged lighting. The cameraman daisy-chained a few extension cords together, stood on a table, and stretched just enough to connect the plug for an overhead light that was supported by exposed pipes. Sands Fish, my lab mate, had recently added a workbench to the space. Though it had a DO NOT TOUCH sign on it, I interpreted that to mean outsiders should not touch. The workbench made of exposed wood chips was topped with a turquoise gridded mat that made a small yet striking stage for my laptop.

On the day of the event, the coded gaze art crew assembled our materials and sorted last-minute posters still wet with black ink, leaving evidence on the fingers of all who helped arrange the show. One poster explained the Aspire Mirror. Another explained how human bias and coded bias connect. My favorite poster was the largest of them all. Printed in white ink was the title "Algorithmic Justice League"—the name I was using to describe the work I was doing. The name follows the "justice league" banner that many others have used since the turn of the twentieth century—decades before DC Comics adopted the term for their fictional worlds—to fight for societal change. In the early twentieth century, civic organizations used the phrase "justice league" in their fight for women's suffrage ("The Equal Justice League of Young Women" [1911]), racial equality and civil rights for African Americans ("Race Justice League" [1923]), and workers' rights ("Justice League" [1914]). Scores of justice-oriented organizations continue to tap into this tradition today. Real-world justice leagues serve as inspiration for the belief that against tyranny, oppression, and erasure, we can choose to resist and offer pathways to liberation. I positioned the emerging work I was doing with the Algorithmic Justice League to follow this banner.

For the occasion, I set up an interactive demonstration of the Upbeat Walls project, where people could attempt to paint walls with their smiles, assuming their faces were detected. As people began to trickle into the exhibition, I started to notice a repeating experience. A fair-skinned person would try the interactive Upbeat Walls and have their face detected and the music start to play. Someone with darker skin would try without luck until they put on the white mask I had put on the table. I overheard a fair-skinned person say, "It works so well for me, I didn't even imagine it wouldn't work for someone else." And someone else, with a darker complexion, commented, "Dang, the machines can't see us either?"

Without seeing someone else struggle with the Upbeat Walls system, the person for whom it had worked just assumed that it worked for everyone. But when they realized that wasn't the case, their new awareness was reinforced by watching a video about the coded gaze. And the person for whom the system failed to work correctly realized that it was the machine, not them, that was at fault. This was not a case of user error. There was something more under the hood, and that was precisely what the whole exhibition was about.

On November 9, 2016, several days after the art show, Donald Trump was elected president of the United States. While some of my childhood friends in Memphis, Tennessee, celebrated, in the streets of Cambridge there was palpable disappointment. Our group—which had been a curiosity in the Future Factory—was now a destination point. People from other lab groups began to make pilgrimages to our Wednesday afternoon meetings, which were generally open to anyone. These visitors were galvanized to think about the role of their technology in today's society, and they came to see what they could learn from our work. I helped pull out more and more folding chairs to hold all the people who joined us. The tables that had once held Ethan up in a moment of despair over funding were now hosting the broader Media Lab community. At our gatherings, students working on new user interfaces, gene editing tools, and unique musical instruments dropped in. There was an undercurrent murmuring, "What is the value of the work I am doing?" and "Can we use the power and access we have at a place like Media Lab to change society?" We were gathering to have a space to voice our fears and our hopes. With our open lab space, Ethan modeled for me what it looked like to hold space for others, no matter where they were on their journey of political awareness, to come and be heard, to ask questions, to express doubts, and to find community.

In an age of rising populism, how could I amplify issues I was seeing with AI, which was now increasingly in the hands of a government that was quickly losing my trust?

I knew I needed to reach more people than had attended the MFA exhibit, though, and I wanted to share my story of coding in a white mask and the increasing use of AI to determine important life decisions like who gets hired or fired. Soon, I had another stage to perform on. I learned there was a TEDxBeaconStreet event taking place in mid-November, and while speakers had already been selected, John Werner, the organizer, was a familiar face around the Media Lab. The main stage schedule was already full, but John offered a slot on the TEDxYouth@BeaconStreet program. It wasn't what I'd hoped for, but it was enough of a crack of the door to get going. I worked on my talk and then I emailed John in an attempt to persuade him to give me a chance on the main stage. He paired me with a speaking coach and resources on how to give a compelling talk. By the time I had refined my talk and attended a few of the practice sessions, John had invited me to the main stage, as long as I felt comfortable given the short time frame. He cautioned that I would have just a few weeks to prepare instead of the normal three months. This was the invitation I was waiting for. I accepted the challenge.

On the day of the event, John greeted me and two friends who had come as emotional support. The day before, they had served as my audience as I practiced. By the time I committed my talk to memory they could repeat some of the parts with me. To add some flair, I decided to repurpose some of the elements of the MFA exhibition by bringing one of the newly minted shields with *AJL* etched on the front on top of a sly smile. Showtime had arrived.

Backstage, a technician outfitted me with a "flesh color" microphone. The pale pink color didn't quite match my gleaming milk chocolate skin. I asked if they had anything darker. They did not. I could hear the murmur of the crowd behind the curtain as the speaker before me walked offstage. With my hand slightly sweating, I gripped a boxy clicker for my slides. Taking a deep breath, I stepped onto the stage. Finding my speaker mark on the floor, I stepped up and began. "An unseen force is rising . . . that I call the coded gaze. It is spreading like a virus."

Slide by slide, the audience leaned in as I explained that from who gets hired or fired to even how much you pay for a product, algorithmic bias is ever present. Algorithmic bias occurs when one group is better served than another by an AI system. If you are denied employment because an AI system screened out candidates that attended women's colleges, you have experienced algorithmic bias.

Glancing down, I saw that the timer had expired. Running into overtime, I rushed to my final slide. "So I invite you to join me in creating a world where technology works for all of us, not just some of us . . . Will you join me in the fight?" The standing ovation was a resounding yes. Momentum was rising.

At this point on my MIT academic path, I began to accept what I could do with the privilege of being at an institution with high visibility. Rather than feeling like I was giving up on my idealism of working on technology to escape painful realities, I could see how my childhood passions could live alongside my solidifying purpose to research harmful discrimination in technology.

My life trajectory and educational opportunities were starting to make more sense to me. I felt emboldened to ask even more uncomfortable questions about the machines that once enamored me.

WITH MY SHIELD ON MY SHOULDER, I RUSHED TO THE MEDIA Lab. Jogging to our great glass elevators, I frantically pushed the sixth-floor button to make it in time for Crit Day. My fellow second-year master's students were exploring so many varied ideas that touched on the thematic trifecta of new technology, aesthetic appeal, and social impact. Nicole L'Huillier, a Chilean musician in the Opera of the Future group, presented on tectonic music to explore new kinds of musical experiences that blended sonic sensations with touch. Stick your head in a box, hold an orb, and feel pressure corresponding to the sound waves that were playing. Udayan Umapathi, who grew up in a small town in India, was from the Tangible Media group. He worked on a project called Droplet IO that allowed the movement of water droplets to be programmed. Using electrical signals, a droplet could be moved precisely to any location and combined with other droplets. From an artistic perspective it could be used to create stunning visualizations, while from a commercial angle the technology was poised to revolutionize microfluidics, the precise manipulation of fluids. Instead of using disposable pipettes that had a negative climate impact, chemical reagents could be mixed with precision, ushering in a new age of digital microfluidics. I was thankful that he presented after me.

In the Camera Culture group, focused on making the invisible visible, Tristen Swedish, who often sported a red beard, was working on eyeSelfie, through which a specialized lens was attached to a smartphone and images were processed with an app that would assist in the diagnosis of eye conditions. The device and accessory would allow for telemedicine in rural areas with limited access to optometrists and ophthalmologists. This work reminded me of the work on trachoma I had done as an undergraduate in Ethiopia.

As each student presented, I made a sketch in my lab notebook to commemorate the day. Then it was my turn. I walked to the front of the crowded room with my shield in hand. The wood-paneled lectern stood like a pedestal directing all eyes to the speaker. I began my presentation with two videos. In the first one I stare into a camera and say, "Hi, camera, can you see my face?" I pause. Nothing. "You can see my friend's face." The video cuts to the face of my friend Mary Maggic, a Chinese American speculative artist. Her face is quickly detected. "What about my face?" The camera returns to my face. I make an exaggerated pout on camera, drawing laughter from the audience. "I have a mask." I put on the white mask, which is immediately detected. "Can you see my mask?" The laughter shifts to audible gasps. On the black screen, three white words linger: "The Coded Gaze."

I then presented another video. In this one I show a person walking in front of a car that is presumed to be self-driving. The car does not slow down and instead collides with the person. I used these videos to tee up the idea that computer vision technology was further infiltrating our lives and that the consequences of not being detected by a computer vision system were not just for chuckles but could in fact be grave. The AI techniques used to detect a face were similar to the techniques used to detect a body. I wanted to show that my focus on faces was only a starting point, and the implications of the work would reach beyond just the face space. An advantage of focusing on an area like computer vision was the ability to demonstrate the kinds of errors being made by a given system. With the white mask example, viewers could see the difference in performance between Mary Maggic and me without a lengthy explanation. Showing and not just telling about computer vision, I reasoned, would allow for powerful depictions of the notion of algorithmic bias and the coded gaze. I call this approach of showing technical failures, to allow others to bear witness to ways technology could be harmful, evocative audits.

The focus of my Media Lab master's work would be "Unmasking Algorithmic Bias." I ended the presentation with a fist pump and raised the AJL shield.

Ethan shouted, "The shield is backwards." I turned it around, ready to field questions.

When the presentations were finished, a woman who had been sitting near the front approached me with a question. It was Cynthia Breazeal. Years after having Cynthia's robot Kismet spark my curiosity, I stood in front of a woman I had long admired, engaging in conversation that had once been but a distant dream.

PART II
CURIOUS CRITIC

CHAPTER 5

DEFAULTS ARE NOT NEUTRAL

Sitting on a purple couch outside Ethan's office, I typed "TED.com" on my keyboard. On the home page was my Beacon Street talk. I took a screenshot to capture the moment, emailed the good news to my friends, and obsessively refreshed the page to check the views. "How I am fighting bias in algorithms" was the title the editors had chosen to introduce my work to the world. The views kept climbing, jumping up in increments of thousands, then ten thousands, then hundred thousands. The video would be viewed more than a million times. People were paying attention. I started reading the comments. Some were kind, but those were not the ones I remembered. From mean comments like "the mask dosent [sic] have to be white you are just ugly" to others doubting that algorithmic bias even existed, my initial elation turned to dread. Was speaking up about these topics going to mean unending verbal abuse and continuous doubts? The spotlight both shines and burns.

The insults were nothing I hadn't heard before, but they still stung. As tempted as I was to comment back, an angry reaction would be counterproductive for people who were genuinely curious or uninformed. "Does this woman not know how Cameras work . . . settings and different lighting has to be used when photographing darker skintones [sic]. Lighter skin tones reflect light better," one person wrote. I wanted to defend my view and show the intellectual case for algorithmic bias. The comments inspired the title "Algorithms aren't racist, your face is just too dark" for an article I wrote shortly after the TED attention. My white mask experience gave me the context that computer vision systems may have some racial bias. My use of bias was based on the idea of disadvantaging or privileging one group or another on the basis of race. Of course people have biases, but as one commenter put it, "There is no bias on math algorithms [sic]." There was a common assumption that these math-based systems make objective decisions; after all, one plus one equals two. Machines were presumed to be free from the societal biases that plague us mortals. My experiences were showing me otherwise. But as I began to discuss algorithmic bias with more people, I was often faced with variations of this hushed and sometimes not so hushed question: Isn't the reason your face was not detected due to a lack of contrast given your dark complexion? (In other words, algorithms aren't racist—your skin is just too dark.)

In the field of computer vision, poor illumination is a major challenge. There are certain instances where we reach the limits of the visible light spectrum. My focus is not on the extreme case as much as the everyday case. The demo on TED.com shows a real-world office environment. My face is visible to a human eye, as is the face of my demonstration partner, Mary Maggic, but the human eye and the visual cortex that processes its input are far more advanced than a humble web camera. Still, even using the web camera, you can see in the demo that my partner's face is not so overexposed as to be undiscernible, nor is my face so underexposed that there is significant information loss.

We cannot fully understand bias in computer vision without a quick look at cameras and imaging technology.

Even though cameras may appear neutral, history reveals another story. The film used in analog cameras was exposed using a special chemical composition to bring out desired colors. To calibrate the cameras to make sure those desired colors were well represented, a standard was created. This standard became known as the Shirley card, which was originally an image of a white woman used to establish the ideal composition and exposure settings. The consequence of calibrating film cameras using a light-skinned woman was that the techniques developed did not work as well for people with darker skin. In fact, it wasn't until furniture and chocolate companies complained that the rich browns of their products were not being well represented that Kodak introduced a new product that better captured a range of browns and dark sepia tones. Separating the appearance of milk chocolate from dark chocolate in advertising had a windfall effect for chocolate-hued individuals. Still, the subsequent digital cameras inherited configurations optimized for light skin, a decision that would translate into a larger set of photos and videos being produced by cameras optimized for only one part of humanity: those with light skin. So when we enter the realm of computer vision, which comes to rely on large datasets, we are dealing with a legacy of cameras and captured images that inherit exclusion. For an image or video, that exclusion can mean that the features of someone with darker skin are less visible even when captured in good lighting conditions.

Default settings are not neutral. They often reflect the coded gaze—the preferences of those who have the power to choose what subjects to focus on. But history has also shown us that alternative systems can be made. In the digital era, the LDK camera series developed by Philips explicitly handled skin tone variation with two chips—one for processing darker tones and another for processing lighter tones. *The Oprah Winfrey Show* used the LDK series for filming because there was an awareness of the need to better expose darker skin, given the show's host and guests.

As I shared the idea of the coded gaze and algorithmic bias more publicly, resistance continued. In spirited emails and online comments, I read some versions of "not EVERYTHING is racist" or "you cannot change the laws of physics." While there are physical limitations, as we saw with the Kodak case, financial incentives can spark ingenuity. In the demo of coding in a white mask, I was also careful to choose an example where my lighter-skinned partner and I were in the same lighting conditions and all our facial features were visible through the camera. For digital cameras, lighting and head pose have an impact on how the pixels appear. I chose the TED demo example precisely to show that yes, even though camera calibration has an impact, it is not the only factor. How we train machines to interpret input is also a key component. Even in the example where through the camera my face was visible to the human eye relying on a well-tuned visual cortex evolved over billions of years, I was not detected. But the white mask I put on was detected. In the exact same conditions, we saw that my light-skinned partner was quickly detected.

Though this example focuses on a face, computer vision can also be applied to attempts to detect cancer or a pedestrian crossing the street. I am less concerned about optimizing computers to detect faces and more interested in understanding how we train machines to see. The white mask demonstration is an entry point to larger conversations about bias in artificial intelligence and the people who can be harmed by these systems.

Given my encounter with the coded gaze, I was particularly interested in computer vision systems for several reasons. The research area that dealt with the ability to detect objects like a face was largely grouped under the domain "computer vision." As a former robotics enthusiast, I was interested in ways to support machines in seeing and interpreting the world. By using computer vision, machines could identify objects or figure out how to move around a space, like a Roomba vacuuming a living room and avoiding collisions with furniture. Not all computer vision tasks, like finding the edges in an image or making a mosaic, are necessarily linked to AI; however, when it came to systems analyzing human faces, AI was being increasingly used.

Between 2010 and 2016, when I went from using my roommate's face to get Simon the robot to play peekaboo to showing a demo of putting on a white mask to have my face detected, there was a massive breakthrough in the world of artificial intelligence. While the old paradigm for programming computers was to provide specific instructions for a task, which can work at lightning pace for something like addition, subtraction, and sorting information, this approach broke down once we pushed computers to do more complicated tasks that might be considered a marker of intelligence beyond the ability to perform mathematical operations.

I think of artificial intelligence as the ongoing quest to give computers the ability to perceive the world (that is, make meaning of visual, aural, and other sensory inputs), to make judgments, to generate creative work, and to give them the ability to communicate with humans. And when I think about computers communicating with humans, that basically means equipping them with the ability to process text or speech, and then training them to respond in a convincing manner. Examples of this ability include asking a voice assistant on a phone for a joke or interacting with a chatbot. Machines can also analyze your behavior and data collected about you to make recommendations that shape our decisions. The decisions can be low-stakes, like Netflix's ability to suggest another film or TV series to binge based on the user's inferred preferences and viewing history. But the decisions can also include more high-stakes situations. For example, AI systems used for employment can recommend a short list of candidates to hire. AI systems used in healthcare can provide recommendations on which patients receive tailored care and which ones do not.[1] Very quickly, we can see how number crunching is not so neutral when those numbers can crunch your life.

IN THE IDEAL CASE, WHICH MOTIVATES MANY RESEARCHERS and engineers, this ongoing quest for artificial intelligence can come together in a complex system such as a self-driving car. The car needs sensory input that can be processed to navigate the world in real time and process visual information: That's another car, that's a tree, that's a human. The car may be equipped with different applications of AI, such as voice recognition so that the car can "listen" for commands like, "Play 'Fear of the Water' by SYML." The car might also provide turn-by-turn navigation. More complicated still, the car may be in a position where life hangs in the balance, where one turn can save a life but end another. Intelligence comes with responsibility. Deriving meaning from complex inputs when there are many potential interpretations makes it very difficult to achieve artificial intelligence by trying to explicitly write code for every potential option.

We saw an example of this difficulty with the breakdown of rule-based expert systems that were once popular in the field of AI. Expert systems were inspired by human experts who have extensive knowledge about a particular area like diagnosing a medical condition or a specific language group like Akan. Expert systems were programmed with a knowledge base of facts and relationships that were explicit for a specific domain, like language translation.[2] To build the knowledge of a translation system, linguists would painstakingly attempt to delineate the rules of different languages to facilitate translation. But the real world never quite follows all the rules. Spoken language and text chat do not exactly follow strict grammar when you need customer service to refund your order.

In 1954, the IBM 701 computer translated basic Russian sentences into English, garnering public interest in machine translation. In a press release, Dr. Dostert, one of the researchers on the project, described the advancement as "a Kitty Hawk of electronic translation."[3] By positioning their effort next to the Wright brothers' first successful flying machine demonstration, the company was not shy about encouraging further investment. Yet enthusiasm had subsided by 1966, when the Automatic Language Processing Advisory Committee (ALPAC) report was released.[4] ALPAC had been commissioned by the National Science Foundation to advise the agency along with the Department of Defense and Central Intelligence Agency on the potential for "mechanical translation of foreign languages." The report showed that the techniques of the day were not on par with the level of quality produced by human translators and recommended investing in ways to help make human translators more efficient as research continued. The ALPAC report showed that the capabilities that had been hoped for, such as mastery of human language, were only just beginning to be explored in the then nascent field of natural language processing (NLP); work on this task continues in NLP to this day.

IN 1956, AT THE FAMOUS DARTMOUTH SUMMER RESEARCH PROJ-ect where the term "artificial intelligence" was introduced, researchers posited that over the course of two months ten men could make significant progress in at least one of the areas they outlined in their proposal, including "how to make machines use language."[5] A decade on, however, the ALPAC report revealed slow progress. An AI winter ensued, where research funding dwindled and graduate students were advised to look at more promising areas of research.

A different approach to the ongoing quest for artificial intelligence was needed. Instead of coding explicit instructions for every conceivable option or pattern and coming up short, a more robust way for machines to perceive and communicate with the world emerged. A baby isn't born knowing how to navigate the world, communicate, or make good decisions. Through observation and imitation, the process of learning unfolds. Taking this process as inspiration, machine learning emerged as one of the leading techniques for artificial intelligence in the early twenty-first century. Instead of machines learning all the rules by explicit instruction, what if we could train them to learn from examples? Compared to rule-based expert systems, machine learning is a less rigid approach to artificial intelligence. Want to have a computer detect a face? Instead of trying to write code to specify all the ways a face could appear in an image, provide a dataset of images with examples of faces. In the realm of computer vision, visual training data can be used to configure a machine to detect a number of objects, including cats, chihuahuas, cupcakes, cultures of cancerous cells, combatants, and civilians. We must keep in mind, however, that these systems do not always get it right. When machine learning systems confuse combatants and civilians instead of mistaking cupcakes for chihuahuas, the consequences are vastly different. One of the ongoing challenges with AI is that the techniques being developed can be optimized for many different types of applications—from benign to lethal.

Since AI can be applied in a range of contexts from cupcakes to combatants, the example applications of the technology shape the public imagination for what is possible, as well as the perceived risks. When companies like Boston Dynamics show their autonomous dog-like quadruped robots doing something cute, like dancing, they mask the ways these systems can be used in the context of military operations or policing. To raise awareness about algorithmic bias, I needed to pick an application that could help people see the risks of AI.

Even though the idea to use machine learning as part of the quest for artificial intelligence has been around since the mid-twentieth century, a few vital components have been provided in the twenty-first century. Machines, unlike humans, often need many examples to learn. In computer vision, an object detection model using machine learning techniques may rely on millions of photos. Before the rise of the internet, access to such a large number of photos was largely impractical. Sites for sharing images, like Flickr, that also include tags provided the treasure trove of labeled data needed. The other necessary component was the ability to store large amounts of data and process that data quickly. Increased computing power and storage with decreased cost helped improve the feasibility and adoption of mobile devices. The proliferation of smartphones in turn allowed more information to be created and shared across the internet. The creators of social media platforms like Facebook and Twitter, and the developers of mobile operating systems like Google's Android, were able to amass large stores of valuable data created by users. Many of us were fueling the advancement of AI unaware.

Advances in the availability of data and increased computing power were important enabling factors. But we still needed a crucial ingredient. How do you get a computer to learn from data? To achieve this feat AI researchers again turned to the brain for inspiration. Our brains are made of neurons networked together in an intricate fashion. As we learn, certain connections are strengthened between our neurons, and others can weaken. No neuron by itself is enough to achieve a complex task like recognizing a face, but by working together, small components can achieve larger tasks. Building on this idea, researchers created artificial neural networks. Instead of neurons and synapses, the artificial neural network contains nodes that are linked to one another in a web of layers. The nodes are inspired by neurons and the links are inspired by synapses. Keep in mind that even though machine learning is inspired by some elements of a biological brain, it does not mean we are creating machines that are sentient or have consciousness. As individuals we are more than our biology.

You can think of the neural network as a pattern recognizer. These networks can take on many kinds of configurations depending on how the nodes are linked with each other and how many layers create the network. You can also think of nodes like marshmallows that can be connected to each other with toothpicks. The architecture of the network is determined by the arrangement of marshmallows and toothpicks. You choose different architectures based on what you want to achieve. Are you making a bridge with multiple levels or a basic V shape? The strength of the connections between linked nodes is defined by weights. If two marshmallows have a strong connection, you can think of that connection as a thicker toothpick or a higher weight. The higher the weight, the stronger the connection. When a neural network is first developed, the weights are not configured to respond to a desired pattern like a face.

To be useful, a neural network must be trained to respond to a specific kind of pattern. The training process of a neural network strengthens some connections and weakens others so the trained neural network model can recognize a pattern. Researchers have developed different kinds of training methods to create machine learning models, which are neural networks configured to recognize a specific pattern. In general, the components of machine learning involve training data, testing data, a neural network to configure, and a learning algorithm to build the experience of the neural network. Remember, an algorithm is a sequence of instructions used to achieve a specific outcome. The goal of a learning algorithm for a neural network is to set the weights between the nodes in the best possible way to recognize a pattern. In the case of detecting an object like a car in an image, the neural network is exposed to many training images that contain a car and the weights are refined many times until the model can consistently detect cars in images it was not previously exposed to.

Like computer vision, the domain of natural language processing has also evolved due to machine learning. Some of the elusive aims from the 1960s, like machine translation of one language into another, are now possible for human languages that are available in large volumes online. Large language models (LLMs), the AI systems that power chatbots, are also trained to analyze patterns. Many LLMs are trained on the text available on the internet, which is to say they ingest vast sums of information from a large portion of what has been made public online. This information includes newspaper articles, scientific papers, standardized test questions and answers, and all of Wikipedia, to name just a few sources that can make them appear extremely knowledgeable. (Imagine taking a test with the benefit of reviewing all the example questions and correct responses that have been posted online and then producing similar responses. I'm not saying you aren't clever, nor am I denying that you have an advantage, but I would be a little less impressed.) These systems ingest not just reputable content, because text on the internet also includes online forums with toxic content, hate speech on social media, and more. LLMs learn the good, the bad, and the ugly.

LLMs go a step further than image classifiers. Instead of being trained to classify a pattern like a face, they are trained to learn language patterns and then reproduce those patterns in convincing ways when prompted. So when you provide a prompt to an LLM like "What indigenous languages are missing on the internet?" you can receive a response that is coherent and grammatically correct. Yet, the response may be unsatisfactory because some missing languages might never have been named on the internet at all. The internet does not represent all human knowledge. The training data provides the neural network with experience that can be used on new data and new prompts. But limited experience has consequences.

A major challenge of neural networks is that during the training process computer scientists do not always know exactly why some weights are strengthened and others are weakened. As a result, current methods do not allow us to explain in full detail how a neural network recognizes a pattern like a face or outputs a response to a prompt. You may hear the term "black box" used to describe AI systems because there are unexplainable components involved. While it is true that parts of the process evade exact explanations, we still have to make sure we closely examine the AI systems being developed. Access to the training data is crucial when we want to have a deeper understanding of the risks posed by an AI system. Unless we know where the data comes from, who collected it, and how it is organized, we cannot know if ethical processes were used. Was the data obtained with consent? What were the working conditions and compensation for the workers who processed the data? These questions go beyond the technical. When I started learning about neural networks, I wanted to know how well they worked in a narrow sense. As a computer scientist, I was trained to focus on the technical, not the ethical. My consciousness about the social implications and the environmental impact of AI began only more than a decade after I first learned how to code.

Returning to the technical, once a system is trained on a collection of data, we assess how well the neural network works at responding to the patterns it has been trained for.

Our understanding of the effectiveness of a neural network is heavily dependent on the data we use to test it. The first neural networks were very simple, with just a couple of layers. Then, as computing capacity increased, researchers could create more complex networks with many additional layers, giving rise to deep learning. Deep learning is a flavor of machine learning that specifically uses deep neural networks, multilayered pattern recognizers inspired by the neural connections of a brain. In this case, imagine a lot of marshmallows and toothpicks. There can be billions of parameters in the systems used to build generative AI products that can create images from a line of text such as "an astronaut riding a horse in space." LLMs can have trillions of parameters. Parameters can include information like how strongly different components of the architecture are connected via weights. Machine learning is also just one approach that can be used for computers to make judgments or generate new content based on prompts. There is more to the story, but this high-level overview is a starting point to uncover how some of the most widely adopted approaches to artificial intelligence become susceptible to harmful discrimination and toxic outputs. Simply because decisions are made by a computer analyzing data does not make them neutral. Neural does not equate to neutral.

Some straightforward decision-making approaches follow explicit rules, such as no one under the age of sixteen being allowed access to content, or a specific score on a risk assessment labeling you as unfit for access to credit. But regardless of the approach, if an automated decision impacts your opportunities and liberties, you must have a voice and a choice in whether and how technology is used.

In my work, I use the coded gaze term as a reminder that the machines we build reflect the priorities, preferences, and even prejudices of those who have the power to shape technology. The coded gaze does not have to be explicit to do the job of oppression. Like systemic forms of oppression, including patriarchy and white supremacy, it is programmed into the fabric of society. Without intervention, those who have held power in the past continue to pass that power to those who are most like them. This does not have to be intentional to have a negative impact. The task ahead of me was to see if I could find compelling evidence showing the coded gaze at work.

CHAPTER 6

FACIAL RECOGNITION TECHNOLOGIES

fter my TED.com debut, people started writing to me about their experiences with facial recognition. One woman told me about the time she was in Las Vegas for vacation and was approached by casino security guards who confronted her for allegedly being a sex worker. Apparently, the casino's camera security systems had mistakenly flagged her as someone else. I also received a handwritten letter from an incarcerated individual pleading that I look into their case because they suspected they were behind bars due to a false facial recognition match. I was horrified and overwhelmed by these stories and the influx of messages about real-world encounters with the coded gaze. With these and many other firsthand accounts about AI failures, the Algorithmic Justice League was starting to feel less like a graduate school project and more like a growing movement.

With the slew of testimonials sent my way, it was clear that faulty technical systems had already contributed to undue scrutiny, suspicion, and, in the case of the inmate who wrote me, jail time. So I started categorizing these examples of machine learning failures. I was surprised to see machine learning could impact my love life. I attempted to sign up for a dating app that appeared to use AI on uploaded photos before allowing entrance. After the system failed to find a face on my first two attempts to upload a profile photo, I returned to my research efforts. I'd already read an account of how failed machine learning could take your freedom away. But to me, it was clear that machine failures don't just mean cases of mistaken identity, detection failures, or other such inaccuracies with the system. Machine decisions can also be applied in a way that fails to adhere to our expectations of fairness.

A Harvard study revealed price differences in the Princeton Review's online SAT tutoring service. After testing thirty-three thousand zip codes, researchers documented that prices for the online service varied by zip code.[1] A Pro Publica investigation uncovered that customers in places with a high density of Asian residents were 1.8 times more likely to receive higher prices for the online tutoring regardless of income.[2] I also thought of the EyeSee mannequin, which was already appearing in stores around the country.[3] Developed by the Italian company Almax SpA, these mannequins were equipped with a camera and paired with software that silently detected shoppers' faces and guessed their age, gender, and race. Unlike the largely visible department store security cameras mounted to walls, these cameras were hidden in plain sight, secretly observing customers at eye level. What would companies do with the demographic information collected by these mannequins? Women already faced a "pink tax," higher pricing for items like pens and razors when they were colored pink as compared to blue or black. Systems designed to assess demographics could use that information in commercial settings to influence which customers received certain promotions and which ones did not.

My task was now to see if the failures I had witnessed or heard about were isolated events or indicative of something more pervasive. I decided to focus my attention on AI systems applied to human faces.

A robust discussion is ongoing about what to call AI systems that analyze human faces. Since AI systems can scan your face with various objectives, specific technical terms are used for different purposes. Researchers, marketers, journalists, and policymakers can use similar-sounding terms to mean different things, thereby causing confusion. Even researchers and companies who work on these systems do not always agree on terminology. When I see the term "facial recognition" in a paper or on a tech website, I first try to understand how it is being defined—if it is defined at all. In 2012, the Federal Trade Commission released a report noting that their use of the term "facial recognition" broadly referred to any technology that is used to extract data from facial images.[4] Throughout the report the term "facial recognition technologies" was used as a catchall phrase. I adopt this phrase because it indicates there are many different types of face-related tasks that machines can perform. But being specific about definitions is more than an academic exercise. The power to define the meaning of terms has a major impact on the reach and effectiveness of legislation and regulation. Companies may also shy away from terms that receive significant public scrutiny and blowback. So instead of saying a system uses facial recognition, a company might say "face matching" to distance themselves from scrutiny. To help deepen understanding about the different kinds of tasks performed by facial recognition technologies, I describe them as explorations of three key questions: Is there a face? What kind of face is this? Have I seen this face before?

Face detection or facial detection describes the task of determining if a video or image contains a human face ("Is there a face?"). Just because a system is designed to perform a certain task does not mean it can do it consistently or that it does not make mistakes. My white mask demo is an example of face detection failure. Not detecting a face that is actually present is known as a false negative. Another way face detection can fail is when a system detects a face that isn't there—a false positive. Artist and technologist Adam Harvey has taken advantage of this vulnerability by producing clothing patterns that can confuse some of these systems into detecting human faces that aren't there.[5] Unsurprisingly, tech marketers don't advertise how their systems might fail, but this gap between what a system is supposed to do in theory versus what happens in the real world can give a false impression about the maturation of certain technologies. Scholars call this the AI functionality fallacy, or the assumption that a system performs the task it was designed to execute as expected.[6] When Google launched Bard, an answer to OpenAI's ChatGPT, the company decided to show off the system's capabilities. In a segment on the television show *60 Minutes,* the Bard system recommended and summarized five books, dazzling the host.[7] After the *60 Minutes* team looked up the books the system recommended, they found out the books did not exist. The titles were made up. A chatbot confidently responding with made-up information is referred to by some AI developers as "hallucination." Author and cultural critic Naomi Klein observes that the term hallucination is a clever way to market product failures. It sounds better than saying the system makes factual mistakes or presents nonsense as facts. These frequent errors are a demonstration of the AI functionality fallacy and a reminder that appearing knowledgeable isn't the same as being factual.

AI systems can also be used to address the question: What *kind* of face is this? What is the perceived gender of the face? What is the perceived age of the face? There are even systems that guess the emotions of a person based on their facial expression. A number of technical terms are thrown around in this arena, though most are self-explanatory. "Age estimation" guesses age. "Gender classification" describes a machine's attempt to determine someone's gender from an image. "Affect recognition" analyzes emotions. And so on. Beneath this technical jargon, however, is a messier, inherently inconvenient truth. For instance, it might be true that the expression on someone's face can at times give you a sense of how they are feeling—that's a big part of how we read nonverbal cues in our real-world interactions—but we also know that just because you can guess something about a person based on what shapes their facial features are making, it doesn't mean your assessment is accurate. We all know you can put on a smile, but does that mean you feel happy inside? There are many examples of categories we use that are not clear-cut, such as gender, which is neither binary nor fixed. Many systems have been developed with the binary gender labels "male" and "female" that erase the existence of trans folks, intersex individuals, or people with more fluid gender identities. When we use machine learning to analyze photos to learn about gender, we teach the machine preexisting cultural norms of gender presentation. You can generalize this lesson beyond gender presentation and see how machines inherit their creators' cultural norms and, by extension, their biases.

Some companies and researchers go as far as claiming that their systems can predict someone's sexual orientation, political affiliation, intelligence, or likelihood of committing a crime based solely on their facial features.[8] I still remember my disbelief when I came across a 2017 study where the authors used images of more than eighteen hundred people to create a classifier to predict criminality based on a face image.[9] I was also alarmed when I read a September 2017 article in *The Economist* about Stanford researchers who made classifiers to categorize someone's sexual orientation based solely on an image of their face. In the article, one of the researchers suggested that with the right dataset they could "spot other intimate traits, such as IQ or political views."[10] Predicting someone's internal state, identity, or future behavior based on facial features alone is unreliable, but the impact can still be consequential. Being labeled criminal or homosexual can lead to harmful discrimination and even death. In December 2020, the International Lesbian, Gay, Bisexual, Trans and Intersex Association released a report documenting that sixty-seven UN-member states criminalize consensual same-sex sexual acts, with six imposing the death penalty.[11] Labels matter, and so we must be extremely skeptical about claims any company or researcher makes about using external features to predict psychological states, innate capabilities, or future behaviors.

"Have I seen this face before?" is the question that researchers categorize under the umbrella of facial recognition as used by industry experts.* Those who work in the biometrics industry do not use the term facial recognition to refer to any kind of data that can be analyzed from a face. Instead, this use of facial recognition is focused on the unique identity of a person. Technically speaking, there are two main types of facial recognition. One is facial *verification,* which is also referred to as one-to-one match-

---

* Some biometric industry practitioners and researchers favor the term "face recognition" over facial recognition. You may also see the term "face verification" instead of facial verification and "face identification" instead of facial identification.

ing. The question the machine is answering is: Does the face presented match the face expected? This is the task that is performed when you use your face to unlock your phone. The other type of facial recognition gets most of the public attention: facial *identification*, which is also known as one-to-many matching. This is when an image of your face—which can be taken from anywhere, ranging from a photo you upload to social media to your appearance on a surveillance camera—is checked against a database of potential matches. The database could be stored by a company like Meta, the owner of Facebook, which has a vast collection of images of people's faces, or it could be owned by a government agency like the FBI, which allows police departments to access those images. The applications for these one-to-many facial recognition systems are vast and limited only by the imagination, and means, of those who wield the power to use them. The pharmacy chain Rite Aid, for instance, installed security cameras equipped with facial recognition capabilities to check faces against a database of prior shoplifters.[12]

Facial recognition systems were being marketed to law enforcement agencies in the United Kingdom as well as the United States. The civil liberties organization Big Brother Watch released a report in 2018 documenting that the United Kingdom's Metropolitan Police Department had piloted facial recognition systems that wrongly matched innocent members of the public with criminal suspects more than 98 percent of the time. The South Wales Police did slightly better with 91 percent false matches. In the process, 2,451 individuals unknowingly had their faces scanned by the department and stored for twelve months.[13] The sooner I could run experiments, the sooner I could gather evidence to help organizations like Big Brother Watch stop harmful use of facial recognition technologies.

When facial recognition works as intended, companies and governments have sophisticated surveillance tools that can be used for social control and exclusion. When facial recognition fails, you can find yourself under investigation for a crime you did not commit, or facing security guards who have "digital evidence" you supposedly resemble a thief, or mistaken for a prostitute in an area where sex work is illegal. The stakes are high either way. And there are more examples: When facial recognition is used to access services like renewing a government passport, failures can also impede your ability to receive vital documentation.[14] When it's used by transportation security, you might be flagged as a terrorist suspect or you might find you are not able to board a train that uses the face to pay or to verify passengers' identity.

Now that you know the differences between facial verification (one-to-one matching) and facial identification (one-to-many matching), you can see why the meaning of the term "facial recognition" needs to be clearly defined when we talk about policy. If we passed a law about facial recognition and defined the term to mean only one-to-one matching, it would not cover instances of facial identification being used for mass surveillance, like during a protest or in a department store. If facial recognition is defined to mean only one-to-many matching, then the law would not cover cases when an asylum seeker or senior citizen is attempting to access government services using facial verification. If we go back to the broad 2012 Federal Trade Commission definition of facial recognition to mean any system that analyzes data from a face, then the law would have broader protections. This broad definition would include gender classification, race classification, or age estimation. Such a law would cover not only cases when your specific identity is captured but also cases in which your demographic or physical attributes could be collected from your face and then be used to violate your civil rights.

A workplace could not adopt a program that systematically screened out job candidates based on their race from face data associated with an online profile link, because it would violate Title VII of the Civil Rights Act in the United States. Title VII states that an employer cannot discriminate against a candidate on the basis of race, sex, or color, among other attributes. Here is where face detection comes in. If a company adopted an AI system that attempted to distinguish real job applicants from fake ones by looking up face images of the people, and the face detection model it used failed on darker-skinned individuals, it would systematically conclude that dark-skinned applicants were fake, putting them at a disadvantage. This action would put the company at risk of violating Title VII. HireVue, a company that claimed to use AI to analyze video of a candidate to infer problem-solving abilities, eventually retired the feature after a complaint from the Electronic Privacy Information Center (EPIC) and an audit from O'Neil Risk Consulting & Algorithmic Auditing (ORCAA). Yet, there are still other companies that use AI systems to analyze videos and faces as part of interview processes. Beyond employment, there are issues in education. E-proctoring companies gained footing during the COVID-19 pandemic when remote learning became a necessity. In an attempt to curb cheating, schools adopted e-proctoring tools to monitor remote learners while they took tests. These companies faced complaints from dark-skinned students who had to set up elaborate lighting contraptions to be seen, or were unable to be verified to log in, or were flagged as cheating. The cheating flag could occur when the system no longer detected a face. However, there are technical reasons a face might not be detected that have nothing to do with cheating but instead indicate a failure of the AI system. Dutch student Robin Pocornie filed a complaint with the Netherlands Institute for Human Rights after enduring remote test conditions that forced her to shine a light on her face to take her exams. White students did not file similar complaints. The institute investigated the complaint and released an interim verdict in her favor, finding that the software used by her university, VU Amsterdam, discriminates against Black students.[15] These systems can also disadvantage individuals with disabilities, as detailed by the Center for Democracy and Technology in a study examining e-proctoring tools:*

> AI video analysis might flag students with attention deficit disorder (ADD) who get up and pace around the room. It could flag students with Tourette's who have motor tics, students with cerebral palsy who have involuntary spasms, or autistic students who flap or rock. It could flag students with dyslexia who read questions out loud, or blind students using screen-reader software that speaks aloud. It could flag students with Crohn's disease or irritable bowel syndrome who need to leave to use the bathroom frequently. It could flag blind or autistic students who have atypical eye movements. Because all of these movements and responses are naturally occurring characteristics of many types of disabilities, there is no way for algorithmic virtual proctoring software to accommodate disabled students. The point is to identify and flag atypical movement, behavior, or communication; disabled people are by definition going to move, behave, and communicate in atypical ways.

* You can read the full report, "How Automated Test Proctoring Software Discriminates Against Disabled Students," on the Center for Democracy and Technology website: cdt.org/insights/how-automated-test-proctoring-software-discriminates-against-disabled-students/.

When companies require individuals to fit a narrow definition of acceptable behavior encoded into a machine learning model, they will reproduce harmful patterns of exclusion and suspicion.

Even before I fully appreciated the harms I suspected and the harms to come, I had my work cut out for me when it came to starting my research—I wanted to direct my work so that it would be put to good use and have some impact in the world. Since facial identification was already being used by law enforcement, and my experience of face detection failure was part of what motivated me to start digging in the first place, I entered the realm of computer vision research. However, because of the wide variety of facial recognition technologies—and the many uses they were already being put to—I realized I needed to focus my efforts and better understand what was already known about the accuracy of these systems.

Coded gaze scorecard showing gender classification, age estimation, and face detection results from five commercial AI products

I started running one of my photos through different systems that analyzed human faces, including online demos from companies like Microsoft, IBM, and Face++ that claimed the ability to analyze faces and guess demographics like age and race. Some of these failed at the basic task of detection. (I was amused to see that in some of my ad hoc tests, the cheeky face on my AJL shield was detected as a face even when my own face remained unseen.) The systems that did detect my face labeled me as male. I also noticed that these systems tended to under-age me. These results were interesting to me, so I decided to take a deeper dive to look at how these systems were analyzing gender and age.

I wondered how these systems would classify the gender of other women with dark skin. I wanted to find more photos to test this question, but the open-source datasets that were available usually contained faces of celebrities, and few of those celebrities were women of color, let alone women with skin as dark as mine. I decided to try the images of women athletes on these systems, and I noticed that some of them were also misgendered. I even went to a website of the MITRE corporation that carried images of deceased inmates available for research purposes. It felt eerie to run the faces. Unlike many of the other datasets I had collected, this one contained a high proportion of Black and Brown faces. In the early days of my explorations in 2016, I was not thinking about consent, or about the ethics of using these images for my research without asking the subjects of these photos. If the image was available in a dataset and my use was for academic research, that was enough for me: This was standard practice in the computer vision research community at the time. But even with this naive attitude, a creeping feeling still came over me as I ran image after image into these cryptic systems.

CHAPTER 7

GUARDIANS ASSEMBLE

I had another problem to solve. No one else in my group at MIT studied computer vision, and I needed to find mentors who could help me grow my technical knowledge to refresh my old skills and expand them. Help came from an unexpected place.

I'd sent an email to the Rhodes Scholars listserv, sharing my budding interest in algorithmic bias. A fellow Rhodes Scholar who saw my message responded and offered to introduce me to Timnit Gebru, a PhD candidate at the revered Stanford Computer Vision Lab (now the Stanford Vision and Learning Lab). Headed by Dr. Fei-Fei Li, this group was famous for creating one of the most important datasets for the computer vision research community, known as ImageNet, a freely available dataset of more than 3 million labeled images. The machine learning approach to artificial intelligence works by training a system on a dataset of examples. To be clear about the kind of pattern being presented to a machine, we provide an example piece of data and a label. So instead of having just a dataset of 3 million images, the images come with classification labels that can indicate if the image shows a car, a cupcake, a cat, and so on. The highly competitive ImageNet Challenge grew from this labeled dataset. Each year, researchers competed to see which of their algorithms could classify images in the dataset with the fewest mistakes and thereby earn bragging rights. Released in 2009, the dataset fueled the advancement of computer vision and the overall field of artificial intelligence. ImageNet showed that strategic data collection, often seen as grunt work and inferior to the development of algorithms, was just as important for advancing artificial intelligence. Data was queen. And it seemed for the time being that size mattered. Large datasets with labels were essential in leading to breakthroughs in the computer vision research community. These labels would also prove useful in tracing out sources of bias.

Talking to Timnit immediately felt like talking to a kindred spirit. We could relate to being first-generation immigrants and we could talk shop about datasets and computer vision research. We could also relate to the challenges of graduate school, from fighting burnout to finding good collaborators. We both rocked our natural hair. She chose the full-out Afro for convenience, while I opted for the faux hawk with flat twists on the side for style. I like to say that our hairstyles reflected our personalities. Timnit was free and practical. I was more measured, but I also had a propensity for flair. I asked her if she could help me brush up on my computer vision skills. She agreed, assigning me the task of creating a tool that would let me easily search the datasets I was interested in exploring. She also sent me Stanford coursework on computer vision to supplement my undergraduate knowledge from Georgia Tech. Soon, with her help, I was up to speed and speaking more fluently about deep learning and computer vision. And while Timnit knew more about computer vision than I did, she was not as familiar with the notion of algorithmic bias. She wanted to learn more, especially as her research interests intersected with computer vision and societal impact.

In addition to Timnit, my de facto computer vision advisor, and Ethan, my official supervisor, MIT required that I have at least two other official readers for my master's thesis committee. I reached out to Professor Mitch Resnick to ask if he would be a committee member. I'd known of Mitch's work for years, having encountered it when I took my first computer science class in high school: He was the force behind LEGO Mindstorms, the programmable LEGO bricks that had allowed me to dabble in robotics engineering as a teenager. He soared over six feet tall, with long elegant limbs, and he reminded me of the Big Friendly Giant from the Roald Dahl book my father had read to me at bedtime.

To complete the committee, I reached beyond the Media Lab and ventured to the MIT Computer Science and Artificial Intelligence Lab (CSAIL), a world-class computer science department housed just across the street. I went in search of Hal Abelson, who was approaching his seventies and had a Yoda-like presence in both stature and demeanor. I walked into his office, whose ceiling spanned two floors—fitting for a man who was considered one of the fathers of computer science education. Hal had shaped the curriculum that formed the foundational understanding of computer science for MIT's top graduates. Before our meeting, Ethan reminded me that Hal was a professor with deep integrity and a reputation for leading with morals even when it was hard or went against the establishment. I hoped his commitment to fighting for justice would resonate with my emerging work on algorithmic bias. I told Hal about the coded gaze and my work with the Algorithmic Justice League to fight this unseen force. Unable to resist the opportunity for some flair, I ended by asking if he would be a guardian of the Algorithmic Justice League.

MY COMMITTEE—THE GUARDIANS OF THE ALGORITHMIC JUS-tice League—had our first meeting in Mitch's LEGO-filled lab. With Yoda, Benjamin Franklin, the Big Friendly Giant, and me huddled around a table, we discussed my exploration into algorithmic bias and the roles each advisor would fulfill along the way.

Hal reminded me that his job was to push me and ask uncomfortable questions, the questions his fellow computer scientists and programmers might be thinking but would stifle in order not to offend. "I may annoy you with my questions, but they will make your work sharper." Ethan insisted that while my work could incorporate artistic components, he wanted to make sure that the technical contribution I made was strong. He also insisted he did not want me to get too comfortable. "I've watched you. The art and creative pieces come easy for you. My job is to push you." Then it was Mitch's turn to speak up. "Work on what excites you!" He wanted me to enjoy the process and pursue the ideas that really moved me.

Mitch was alluding to a learning approach he'd outlined in his book *Lifelong Kindergarten*, where creative learning is supported by four *p*'s: projects, passion, peers, and play. "Play" was what he was invoking in this moment, the idea of keeping an open and curious spirit, allowing for happy accidents and unanticipated pathways to emerge. This goes hand in hand with the idea of hard fun, a term conceptualized by the mathematician and AI pioneer Seymour Papert. Hard fun is what's happening when we willingly take on difficult subjects, or work through mundane tasks, because we're working on projects that impassion and excite us. Hard fun is what my high school classmates and I were experiencing playing with Mitch's LEGO Mindstorm kits during lunch period, and what we experienced in Jill Connell's classroom at Cordova High, when we were wading into the technically challenging, yet alluring, world of computer science. Or in my science fabrication class where, ignoring the siren song of a beautiful New England autumn, I'd spend entire days beneath the fluorescent lights of a lab space building my Aspire Mirror. As Mitch suggests, a spirit of play and a commitment to hard fun—especially in computer science and engineering, which some perceive as dry and extremely challenging—can lead to big breakthroughs.

In addition to my advisors, I also had peers like Timnit, whom I looked up to and could rely on as a constant source of support. I felt Timnit understood my research in ways that my committee members did not. This understanding came not just because of our shared experience as young Black women navigating academia, but also because she had expertise in computer vision that my committee lacked. With her, the idea that my skin and gender might have something to do with the errors I had recorded so far was readily received from a technical standpoint and as a person of color. My committee wanted much more evidence that something more was happening. Timnit's mentorship encouraged me to pursue my hunches about algorithmic bias. After some initial hesitation, I set out to collect the necessary data to answer my basic research question: Do AI systems that classify gender based on an image of a face perform differently based on the skin type of a face?

With my guardians assembled and my growing friendship with Timnit, the Algorithmic Justice League was becoming more than an idea. I imagined AJL becoming a network of individuals from different backgrounds working together to uncover what ailed artificial intelligence so we could create better systems that prevented harms instead of perpetuating them. While we were at it, I wanted to maintain a playful attitude that kept the work inviting to outsiders and helped me go through the grind of day-to-day research. The work ahead would be tedious, but the results could transform the trajectory of AI.

CHAPTER 8

POWER SHADOWS

The hard fun began in earnest. Developing a system to detect, classify, or recognize a face is half the battle. You collect a dataset of faces to do the training. The other part of the battle is evaluating how well that system works.

For researchers, the typical practice is to select a benchmark, a dataset used as the standard against which newly developed systems can be tested. And in any given research community, there's usually consensus among researchers regarding which benchmarks are especially rigorous. For instance, in the early years of developing one-to-one facial recognition systems that could later be used to unlock devices like smartphones, a dataset called Labeled Faces in the Wild (LFW)—a set of more than thirteen thousand images of nearly fifty-eight hundred individuals—became the gold standard benchmark.* One of the major issues with prior benchmarks was that most of the photos in these datasets were taken in controlled environments. Imagine what happens when you sit in front of a camera for a professional portrait: The lighting is just right, the camera is pointed straight at you, and the photographer directs your posture so that your face is clearly captured. Early facial recognition systems could perform well on these benchmarks but they suffered a performance drop once they had to deal with real-world conditions. Labeled Faces in the Wild was a big step forward because it was composed of images collected "in the wild"—that is, images not taken in studio conditions—and therefore pushed researchers to create systems that would theoretically perform better in the real world.[1]

AS RESEARCHERS DEVELOPED NEW APPROACHES FOR ONE-TO-one facial recognition, their peers would judge the state of the art based on its performance against LFW. For a while the accuracy plateaued at below 80 percent.

Then in 2014, a major breakthrough came into play when researchers at Facebook released a paper called "DeepFace" that reported performance on the gold standard LFW benchmark of 97.35 percent, breaking the plateau. There was much rejoicing in the artificial intelligence community because it showed the promise of using deep learning for facial recognition. It's no coincidence that Facebook was able to do impactful research on facial

* The dataset contained 13,233 images of 5,749 individuals. It was intended to help develop research methods, but was ultimately appropriated by some companies to promote the accuracy of their products. The creators eventually added a disclaimer to dissuade companies from using the benchmark results for marketing purposes.

recognition: The users on their social media platform had uploaded personal images containing faces, and many images were tagged to other Facebook users; those networks provided a vast dataset for the company to explore for research purposes.

Reporters received the "DeepFace" paper with enthusiasm, hyping it up to the general public with headlines like "Facebook's DeepFace Project Nears Human Accuracy in Identifying Faces."* Reading this paper, and others that soon followed in its tracks, I could see why there was growing confidence in facial recognition systems. With research coming out of influential companies like Facebook, Google, and Microsoft, coupled with research by some of the best computer vision labs in the world, showing high accuracy rates on the gold standard, it wasn't a surprise that people assumed the technology was mature enough to start using in the real world. This sense of optimism was also bolstered by government reports. For instance, the National Institute for Standards and Technology (NIST) also had benchmarks for a range of technologies focused on biometrics like fingerprints and faces. Companies that developed biometric technology for law enforcement submitted their systems to NIST for testing. NIST benchmarks went beyond facial verification and included tasks like gender classification and facial identification. The government studies from 2010 to 2014 showed overall improvement in the performance of facial identification systems.[2] In research circles, these

---

* This headline is a good example of how terminology about facial recognition in the press can differ from academic terminology. Though the headline describes the "DeepFace" project as "identifying faces," this was not an example of facial identification (the one-to-many subtype of facial recognition). Instead the researchers conducted facial verification (the one-to-one subtype of facial recognition). The full title of the paper is "DeepFace: Closing the Gap to Human-Level Performance in Face Verification," by Yaniv Taigman, Ming Yang, Marc'Aurelio Ranzato, and Lior Wolf: research.facebook.com/publications/deepface-closing-the-gap-to-human-level-performance-in-face-verification/.

promising results increased confidence in the use of facial recognition in the real world.

However, just because a benchmark is adopted and becomes the status quo doesn't mean it should go unquestioned. Despite the advances I read about, my own experience of coding in a white mask made me skeptical. At the very least, I wanted to take a deeper look at the underlying details. How golden were these standards in the first place—what standards were *they* held up to before researchers decided their value to the research community? For the most part I found little research that looked into the demographics of who was included in datasets like LFW. This lack could be partially explained by the fact that these images were scraped off the internet out of convenience. And then labeling and analyzing datasets also took time and resources. So it was common practice to collect a large dataset—like a collection of photos containing human faces—and know only that the dataset contained X number of faces without necessarily knowing the exact demographic breakdown of the faces. If I *did* find research that reported on the demographic composition of datasets, the picture was stark. In 2014 Hu Han and Anil Jain examined the demographic composition of LFW; they found that the database of images contained 77.5 percent male-labeled faces and 83.5 percent faces labeled white.[3] The gold standard for facial recognition, it turned out, was heavily skewed. I started calling these "pale male datasets."

THE HAN AND JAIN PAPER INSPIRED ME TO LOOK CLOSELY AT other benchmarks. IMDB-Wiki was another public face dataset that contained images of celebrities and public figures.[4] It made sense that benchmarks like IMDB-Wiki relied on photos of famous figures: There's not only a plethora of images available online, but the images can also be readily labeled with identifying information, which is helpful when training machines to recognize faces. As a result, however, these datasets reflected the demographic makeup of Hollywood, not the rest of the population—and skewed mostly young, white, and male.

I then turned my attention to the datasets that were coming out of NIST, which has two types of datasets, broadly speaking: sequestered data that is used to test systems internally, and public datasets released to help advance research in the field.

At the time I was doing my MIT master's thesis, NIST had released a public dataset of faces called IJB-A. The aim of this dataset was to provide increased diversity—specifically, geographic diversity—and also overcome another challenge with developing benchmarks: face detection failure. To automatically collect photos of images online, instead of having a human search for photos one by one and inspect the image for a face, researchers wrote code that exploited search engines to conduct image searches. Since the researchers wanted only images containing faces and not all the other kinds of images one might find on the internet, they would often include code for automated face detection to filter down the images. As a result, the composition of a dataset collected using this method would be highly dependent on the quality of the face detection code used for the job. Researchers found that while they could collect massive amounts of face data without having to obtain consent, they were missing useful data that contained faces missed by the face detector.

My example of coding in a white mask raised the question of whether the skew away from darker skin was also compounded by the face data collection methods themselves. To overcome this challenge, IJB-A was collected without a face detector and was curated by humans to further increase the difficulty level. Given the aim of demographic diversity with this dataset, I decided to label it to see its composition. Even with IJB-A, developed in 2015, eight years after LFW, I found that the dataset was 75.4 percent male and also 79.6 percent lighter-skinned individuals. This concerned me: Not only did we have a research gold standard that was heavily skewed, but the government agency NIST, which had set out to collect intentionally diverse datasets, also had a dataset that underrepresented women and people of color.

I took my analysis further and did an intersectional analysis informed by the work of Kimberlé Crenshaw, a leading legal scholar. Her path-breaking work on antidiscrimination law in the United States revealed the limitation of looking at single-axis analysis such as gender or race when investigating discrimination cases. She showed that individuals like women of color who faced multiple forms of intersecting discrimination were being overlooked by the way the laws were written.[5] If you were discriminated against because you were a woman or Asian or Black there could be redress, but the bounds of the law didn't allow room for claiming discrimination on two fronts. To illustrate, the Equal Employment Opportunity Commission (EEOC) had a four-fifths rule stating that if you can show that the minority group is getting less than 80 percent of similar opportunities as compared to the majority group, there was a basis for a case. The problem for a Black woman would be that for all women at a company or for all Black people at a company the numbers in aggregate might well be lawful and as long as white women were getting opportunities and Black men were getting opportunities there wasn't a basis to claim gender or race discrimination at the company. However, this meant that if Black women or Asian women were not getting those opportunities, they didn't have a legal basis to seek redress because the employer could say they hire women and they hire Black people or Asian people. But as writer-activist Audre Lorde reminds us, people do not live single-issue lives.

From my vantage point, it appeared that the machine learning community was not yet applying these insights from anti-discrimination scholarship. The focus on benchmark accuracy was oftentimes just looking at one metric: overall performance. I wondered, What would it look like if we started to tease performance apart across multiple axes of analysis? What would we learn about the capabilities and limitations of a system? How far would we apply a benchmark if we looked at which groups were included and, just as important, which groups were excluded?

Encountering Crenshaw's intersectionality work—in a seemingly unrelated field from my own—opened a pathway to ask deeper questions about my research and about AI more broadly. My undergraduate computer science training had prepared me to look under the hood of machine learning systems, and my personal experiences with the coded gaze galvanized my curiosity, but it was the scholarship of Black women scholars I encountered in graduate school that would give me the language to articulate what I was observing in AI. I applied this idea of intersectionality—or analyzing across multiple axes of identity—to my evaluation of the NIST benchmark. The intersectional analysis was a revelation. When I looked at the composition of the government dataset not just by gender or skin type individually but also by the intersections of multiple factors, I found that lighter-skinned males made up 59.4 percent of the entire benchmark and women of color were only 4.4 percent of the benchmark.[6] Even if a system failed on *all* women of color in the dataset, it could achieve an accuracy rate of 95.6 percent on the entire benchmark and be deemed suitable for the real world. Thinking beyond faces, deep learning techniques were being applied to systems trained to detect skin cancer or to detect pedestrians to be used in self-driving cars. If those datasets were also skewed, it could mean AI cancer detectors would not work well for groups of people underrepresented in the dataset. It would mean automated vehicles would be more likely to crash into some groups of people than others.

I began to better understand why what I was reading in research papers and what I was experiencing were at odds. The benchmarks themselves masked potential bias because they lacked representation. The gold standards were pyrite, fool's gold, presenting glittering accuracy in their numbers while not being composed of a structure representative of all of humanity. Yet, they were accepted by experts as the status quo.

I was left with one major lesson: Always question the so-called gold standards. Just like the standard Shirley cards that were used for calibrating film-based photography might seem neutral or untouchable, standards used in AI may appear to be off-limits for questioning when we assume the experts have done a thorough job. This is not always the case, particularly when the experts do not reflect the rest of society. But design is not destiny. I knew I had to try to change the system.

WHEN MACHINE LEARNING IS USED TO DIAGNOSE MEDICAL conditions, to inform hiring decisions, or even to detect hate speech, we must keep in mind that the past dwells in our data. In the case of hiring, Amazon learned this lesson when it created a model to screen résumés.[7] The model was trained on data of prior successful employees who had been selected by humans, so the prior choices of human decision-makers then became the basis on which the system was trained. Internal tests revealed that the model was screening out résumés that contained the word "women" or women-associated colleges. The system had learned that the prior candidates deemed successful were predominantly male. Past hiring practices and centuries of denying women the right to education coupled with the challenges faced once entering higher education made it especially difficult to penetrate male-dominated fields. Faithful to the data the model was trained on, it filtered out résumés indicating a candidate was a woman. This was the by-product of prior human decisions that favored men. At Amazon, the initial system was not adopted after the engineers were unable to take out the gender bias. The choice to stop is a viable and necessary option.

The face datasets I examined revealed data that was not representative of society. The example of the Amazon hiring model illustrates what happens when data does indeed reflect the assumptions of society. Their model reflected *power shadows*. Power shadows are cast when the biases or systemic exclusion of a society are reflected in the data.

Seeing the major skews toward lighter-skinned individuals and men in the face datasets motivated me to understand why these biases happened. How were these datasets collected in the first place? When I looked at the government benchmark as a starting point, answers started to emerge. To attempt to overcome privacy issues, the researchers chose to focus on public figures, who, by the nature of their jobs in society, often as public servants, had a level of visibility that made information about their demographic details public knowledge. While using public figures could potentially overcome some privacy concerns, the choice itself came embedded with power shadows. Who holds political office? It is no surprise that around the world men have historically held political power, and to this day we see the patriarchy at play when it comes to leadership and decision-making. At the time I conducted my research, UN Women released a chart showing the gender gap in representation for women in parliaments. This analysis revealed that on average men made up 76.7 percent of parliament members.[8] So when creating a dataset based on parliament members, the shadow of the patriarchy already lingers. While that could in part lend a plausible explanation to the male skew, I also wanted to gain more insight into the disproportionate representation of lighter-skinned individuals. The work of Nina Jablonski on skin distribution around the world shows the majority of the world's populations have skin that would be classified on the darker end of most skin classification scales. Returning to the government IJB-A dataset that was created to have the widest geographic diversity of any face dataset, how was it that the dataset still was more than 80 percent lighter-skinned individuals?[9]

When we look at who holds power around the world we see the impact of colonialism and colorism that derives from the power shadow of white supremacy. Formerly colonized nations when they became independent still inherited the power structure of colonialism. White settlers and their offspring were often lighter than the indigenous people of a land or darker African enslaved people brought into colonized countries. When I started looking at the composition of parliaments around the world, I saw this impact. In South Africa, despite the population being classified as 80.8 percent Black, 8.7 percent colored, and 2.6 percent Asian, around 20 percent of the parliamentarians would be classified as white.[10]

Stepping beyond a colonial past does not decolonize the mind. White supremacy as a cultural instrument, like the white gaze, defines who is worthy of attention and what is considered beautiful or desirable. Colorism is a stepchild of white supremacy that is seldom discussed. Colorism operates by assigning high social value and economic status based literally on the color of someone's skin so that even if two people are grouped in the same race, the person with lighter skin is treated more favorably. We can see this in Hollywood and Bollywood. India with its vast diversity of skin types has an entertainment and beauty industry that elevates light-skinned actors and actresses. Women are judged on their beauty, and the standard of beauty is predicated on proximity to fair skin. Beyond beauty, lighter skin is also associated with having more intelligence in societies touched by white supremacy. Hollywood has long favored white actors, and when it began to open up slightly, lead roles for diverse cast members also skewed to the lighter hue. This is not to say that at the time I was doing this research there were no dark-skinned individuals who had gained fame or were positioned as intelligent. But the fact that they were the exception and not the norm is the point.

Going back to face datasets, we need to also keep in mind how the images are collected. When a group like elected politicians is chosen as a target dataset, the images that are collected are based on videos and photographs taken of the individuals. Here again we can see how the shadow of white supremacy grows. Which representatives are more likely to have images and videos available online? If you make a requirement that to be included in the dataset you need at least ten images or video clips, the representatives that receive more media attention are going to have an advantage. Even if you do not filter using automated methods like face detection, which has been shown to fail more often on darker-skinned faces, the availability of images based on media attention will still favor lighter-skinned individuals. Despite the intention to create a more diverse dataset with inclusion of representatives from all around the world, the government dataset was heavily male and heavily pale, inheriting the power shadows of patriarchy and white supremacy. These are not the only kinds of power shadows to contend with. For example, ableism, which privileges able-bodied individuals, is another kind of power shadow often lurking in datasets, particularly those used for computer vision. For pedestrian tracking datasets, few have data that specifically include individuals who use assistive devices. Just as the past dwells in our data, so too do power shadows that show existing social hierarchies on the basis of race, gender, ability, and more. Relying on convenient data collection methods by collecting what is most popular and most readily available will reflect existing power structures.

Diving into my study of facial recognition technologies, I could now understand how, despite all the technical progress brought on by the success of deep learning, I found myself coding in whiteface at MIT. The existing gold standards did not represent the full sepia spectrum of humanity. Skewed gold standard benchmark datasets led to a false sense of universal progress based on assessing the performance of facial recognition technologies on only a small segment of humanity. Unaltered data collection methods that rely on public figures inherited power shadows that led to overrepresentation of men and lighter-skinned individuals. To overcome power shadows, we must be aware of them. We must also be intentional in our approach to developing technology that relies on data. The status quo fell far too short. I would need to show new ways of constructing benchmark datasets and more in-depth approaches to analyzing the performance of facial recognition technologies. By showing these limitations, could I push for a new normal?

PART III
RISING
RESEARCHER

CHAPTER 9

CRAWLING THROUGH DATA

Early spring blossoms were attempting to coax Cambridge out of stubborn winter, and I was now in my second year at MIT. The consequence of switching my research direction in the fall of 2016 meant that I would have just a few months to conduct my algorithmic bias experiments and a few more to write a thesis about my findings and their implications. Despite being only twenty-seven years old in the spring of 2017, I did not feel time was on my side—there was so much to do. My experience of coding in a white mask demonstrated an example of face detection failure, but I wanted my MIT research to expand beyond detection to show yet another area in the study of the face that needed attention. I narrowed my research to systems that focused on guessing attributes of a face, such as gender, race, and age. Unlike race and age, which could have many different categories, almost all the gender classification systems that existed then provided only two options: male and female. Though gender is not binary, the use of only two gender options by most AI systems made focusing on binary gender classification a more convenient choice. Still, I explored other options. Robust race classification and age estimation would require more than two groups. I experimented with having workers on Amazon Mechanical Turk (a platform that allowed researchers to put out low-priced micro tasks for crowdsource workers to complete) assign age, gender, and race labels to images from an existing face dataset. The same faces would be shown to multiple workers known as turkers, and I would examine the labels. When it came to age, having turkers guess a defined range instead of a specific age produced more consistent results. When it came to guessing race, I first used categories from the U.S. Census and left open an "other" category. The results of that category provided the most insights. It became clear that the Census categories were ill-equipped to deal with people the turkers perceived as South Asian, Southeast Asian, Middle Eastern, or of mixed race. (Because the faces shown were of public figures, the turkers had more information at their disposal than just the image being shown.) Of the three categories to label, gender yielded the most consistent results, tipping the balance toward focusing on gender classification.

These explorations show how intimately involved humans are in the process of developing automated systems. Examining the different labels turkers gave to the same face made me see the extent of guesswork that went into attempting to categorize perceived race. After the turkers' experiments, I started using "perceived race" instead of "race" when talking about classification. Setting up the micro tasks also gave me power and privileged my perspectives. It was my human choice to select categories for classification that the turkers were then boxed into attempting to fit.

My own choice of classification categories was informed by how others had grouped people in the past. I looked to existing systems as a starting place, such as the U.S. Census race categories, which have evolved over the centuries and reflected the social, political, and economic context of the day. For example, the U.S. Census, which started in 1790, did not allow people to categorize themselves into multiple racial groups until 2000. It was in 1960 when census takers could first self-identify how they fit into the given options. The categories themselves evolved, with enslaved people becoming "colored," then "Black," then "Negro." The term "African American" debuted in the 2020 census.[1] None of this labeling felt precise, and my choices, just like the census labels, were not neutral. They were subject to the time period and shaped by those who held decision-making power.

After deciding to focus on gender classification for the sake of the technical simplicity of binary classification, I still had to deal with the notion of race. My face not being detected in the first place, I reasoned, had more to do with being dark-skinned than with being a woman. I didn't want to test gender classification for the sake of it. I wanted to see if these systems changed performance for different groups of people. I needed to factor in more than gender categories in my experiments, and so I began an unexpected exploration into an area of study known as ethnic enumeration. I learned that across the world the rationale for categorizing people by race and ethnicity differed, and even the use of the terms race, ethnicity, or color to define the categorization varied. For instance, to uphold apartheid in South Africa, racial classifications were specifically and explicitly linked to economic, social, and political relations. Being classified white, colored, or Black made a huge determination in an individual's life opportunities.

When I visited Cape Town, South Africa, in 2019 during a tech conference, I went to the "Classification Building," where people could go to have their hair and even their most private parts examined to determine race. The example of Sandra Liang, born to white Afrikaans parents but presenting in a way classified as colored, is an example that reveals how race is constructed. It was thus possible for parents classified white to birth a daughter classified colored who was ostracized by the white community and eventually found refuge in a township. In places like Canada, the term "visible minority" is used, a term that acknowledges that outward appearance is what is used to make assumptions about race that hold real social consequences. In the Canadian case, and in places like the United Kingdom and the United States, racial categorization is used by the government to better understand where to allocate resources and support minority groups or those who might face discrimination. In the U.S. case, I found a contradiction in this stated goal and the existing census classification, which included Middle Easterners in the group of white. The treatment of and discrimination against individuals perceived as white and those perceived as Middle Eastern differed, particularly after the September 11th World Trade Center attack. This would hardly be the only contradiction to be uncovered.

I was also surprised to learn about the wide variety of ethnic groups in Europe, as I had been socialized to use the broad label of white for people with European roots. I found that for European countries like France and Germany, forgoing ethnic enumeration for the aim of building a national identity—despite a large diversity of ethnicities with their own cultural customs and oftentimes manner of speaking a shared language—was a deliberate act. Reading this reminded me of a trip I took to Scotland with my friend Alan from Taiwan, who opened my eyes to the large number of different ethnic populations in and around China. I shared with him that there were more than forty indigenous languages spoken in Ghana and a similar number of tribal groups; I told him my mom had a pretty good eye for distinguishing tribal membership, at least from my vantage point. It also reminded me of my time spent in Zambia on a Fulbright fellowship, where locals often asked me my tribal affiliation, assuming I might be from the region. Also in Zambia my students pointed out that I have a West African smile. I didn't know there was such a thing until I did a Google search, which to their delight in fact showed I did have what they called a West African smile. I shared these experiences with Alan as we walked behind two women discussing a new article that claimed there were distinct DNA differences between Scotsmen and Englishmen. Alan, whose partner was a Frenchman, and I also talked about our amusement at seeing French and English animosity as we navigated Oxford University during our Rhodes Scholarships. To us they all looked "white," just as to me Alan looked "Asian," and depending on the part of the world I was in I looked "Black," or for people with more experience with "Black" faces, I looked "West African." No one ever guessed I was a mixture of Ashanti and Dagua tribes.

The mixture of national approaches to ethnic enumeration, regional specificity, and outward appearance plays a role in race and ethnic classification, which is anything but fixed. Living in different parts of the world and in between cultures also gave me an appreciation for how phenotypic perceptions changed. In Zambia and Ghana, being Black made me part of the majority. My parents grew up without race being a primary social issue. When we moved to the United States, it took some time before they would link negative experiences to race. If I received bad treatment, they wouldn't immediately assume it was because of race, but instead they would want to know the details of the situation. I grew up in Oxford, Mississippi, so my racial consciousness was shaped by the context of Black people being minorities and being stereotyped. When my elementary school classmate Billy invited the white kids in our class to his birthday party, but excluded me, I was pretty sure it was because I was Black. But it could be that he didn't like me, or some combination of the two. I wanted to leave these memories behind when I got into a technical field like computer science.

At first I thought my research would be deeply focused on technical issues. Digging deeper made me see that any technology involved with classifying people by necessity would be shaped by subjective human choices. The act itself is not neutral because classification systems do not come out of nowhere. This is what is meant by the term sociotechnical research, which emphasizes that you cannot study machines created to analyze humans without also considering the social conditions and power relations involved.

Despite these complexities, some researchers nonetheless attempted to create machine learning models to guess race and/or ethnicity, oftentimes not distinguishing the two. Some studies were so crude it was almost comical—their labels included "white" and "non-white." Others tried to borrow from existing classification systems and used labels like "caucasoid" and "negroid," classifications that have roots in eugenics and scientific racism. I even found a website called Ethnic Celebrities and created a system that collected the images with a combination of race and ethnic description of all the celebrities. I went as far as finding one of the most comprehensive ethnic classification systems from the Australian Bureau of Statistics, which enumerates more than 270 different cultural and ethnic groups.[2] Despite my efforts to try to make some technical distinctions, my attempt to label celebrities by ethnicity broke down, particularly when dealing with people who were multiracial. Then there was the challenge of trying to tease out how people identified as Hispanic should be classified. On the U.S. Census, Hispanic is the only ethnicity group included that can be applied to different race groups (that is, "White, Hispanic" or "Black, Hispanic"). Knowing I wasn't the first to attempt this work of defining racial classification for face datasets, I turned to NIST. They hadn't fared much better. Some of their studies used "white" and "black" labels, while more comprehensive studies eschewed racial categories and focused on nationality. As a marker for performance on different populations, however, nationality left much to be desired. In countries with large racial and ethnic diversity like Brazil or the United States, the overall country-level performance did not reveal much about the differences between racial or ethnic groups.

After a few weeks of trying to disentangle and arrange this evolving social construct of race, I tried another perspective. Inasmuch as I was looking at gender classification, my aim was not to see if I could come up with a better race or ethnic classification system but to see if someone's appearance made a difference in the accuracy of gender classification. I decided instead of looking at race I wanted to look for a more objective measure, which is when I started to focus on not just demographic attributes like gender and race, but also phenotypic attributes, namely the color of someone's skin. Since face-based gender classification using photos relied on imaging technology, and since skin responds to light, focusing on the color of skin seemed to be a way to be more specific and objective. So, I began looking at ways people have classified skin.

While I thought I was exiting the complications of race classification, I soon found that the Felix von Luschan scale, used by anthropologists, was also employed in ways that supported scientific racism. Along the way I switched to exploring how dermatologists, not anthropologists, look at skin. Instead of just looking at skin color, which can change when exposed to sunlight, dermatologists look at skin response to UV radiation—thus skin type. Skin type and skin color are related but not the same. For example, tanning in the summertime can change your skin color without changing your skin type. Focusing now on skin type, I learned of a more scientific measure called the Fitzpatrick skin phototype scale.[3] When it was first developed in 1975, by Harvard scientist Thomas B. Fitzpatrick, it had four categories. The first three categories were different ways skin that was often classified as white in the United States reacted to sunlight, and the fourth category was for everyone else (the majority of the world). In the eighties the fourth category for "nonwhite" was further expanded to three more. It still wasn't exactly a balanced scale, but it was more scientifically rooted and less complicated than the thirty-six-point Felix von Luschan scale.[4] So I made the choice to use phenotypic classification based on skin type for the dataset I would need to create for my research. But I was dreading explicitly labeling strangers by the appearance of their skin.

Looking at how machine learning models are being developed, we see the impact of what I first heard Kate Crawford term the "politics of classification." Having the ability to define classification systems is in itself a power. The choices that are made are influenced by cultural, political, and economic factors, and while these classifications don't have to be based on definite distinctions, they still have an impact on individual lives and societal attitudes. Despite the power shadows inherent in classification systems, often those systems go unchallenged. Instead, they are

used as shorthand. Morgan Klaus Scheuerman and their colleagues have done tremendous scholarship on how gender classification systems lead to erasure, reification of social constructs,
and the reinforcement of gender stereotypes.[5] In my research, I
had to contend with understanding the limitations of classification systems we use as well as acknowledging those systems'
utility in showing discrimination, unfairness, or inequitable treatment. Yes, the classification systems were problematic, but in
using them I could show their limitations as well as the limitations of the other systems I examined. To contend with those
limitations, it is important to call out the assumptions of the classifications that are being used so there isn't a universal acceptance
that these are the ones that must be used or that they do not have
problems. So, while acknowledging that gender isn't binary and
that the Fitzpatrick scale, the most scientific and least racist skin
type scale I surfaced, was heavily skewed toward lighter skin, I
nonetheless selected two classification systems that would help
me show why we must question classifications made by machine
learning models in the first place. For me, this started with binary
gender classification. Crawling through classification systems and
determining the labels to be used are important parts of the process, but these labels still need to be applied to data.

After going through the maze of classification systems, encountering the limitations of binary gender classification, and
switching to skin type classification instead of race-based classification, it was time for the next hurdle: collecting my own dataset.
For researchers, if it is possible to use prior work, doing so is the
default approach. Before embarking on making my own dataset I
looked to see what datasets were available for gender classification. I found the Adience dataset, which had been specifically
made to use with research studies on gender classification. As one

might expect, this dataset had near gender balance, with 48 percent male faces and 52 percent female faces, but it was still overwhelmingly pale, with 86 percent lighter-skinned individuals.[6] This skew toward light skin even with a gender-focused dataset prevented me from using it for my research because I was interested in how systems would perform on darker-skinned individuals. After considering other datasets, I was sufficiently convinced that I would not be reinventing the wheel by creating a new dataset.

It is one thing to critique other datasets and point out the shortcomings of prior research; it's another to try to create one for yourself. My exploration of the construction of datasets using convenient methods showed that the existing approaches were lacking. Power shadows made it more likely that using automated methods of data collection or celebrity-based methods of data collection would result in heavily skewed data. I had to develop another approach.

The United Nations provided a helpful starting place. A UN Women report on gender representation revealed that on average men made up around 77 percent of parliamentarians around the globe. On the Inter-Parliamentary Union website, I uncovered a chart that would prove vital to my dataset construction efforts. The chart ranked each UN member country by their representation of women in parliament. Rwanda led the world with 61 percent representation of women, which could be attributed to systemic changes in the law that required gender parity in political representation. In the top ten were two other African nations, including Senegal and South Africa, which was tied with Finland for ninth place with 42 percent.[7] The high rank of these African countries that had better representation of women in power than their global peers gave me a starting point to find publicly available images of public figures who were women and had a higher chance of having skin type on the darker end of the Fitzpatrick scale. I still would have to dig into the numbers, because assumption is not fact.

Also in the top ten nations, I found progressive Nordic countries whose egalitarian ideals seemed to be somewhat reflected when looking at the balance of women with parliament seats. It also helped that in Finland, Iceland, and Sweden, fair-skinned people on the opposite (lightest) end of the Fitzpatrick skin type scale would be found in high numbers. While there were countries in Central America that also made it into the top ten, when I examined the parliamentarian representation it seemed their skin types would fall more in the middle range of the Fitzpatrick scale or into categories that were not well accounted for by the scale, so I decided to focus on people who would fall into the first two categories of the scale or the last two categories. (Out of curiosity I looked at where the United States ranked. I scrolled for a while. The United States, with 19 percent, was number 100 out of 193 slots.)

Having narrowed down the choices, I started data collection on three of the high-ranking African parliaments and three of the high-ranking European parliaments. At this point my curiosity got the best of me so I also examined the parliaments of many other countries, including Singapore, India, Brazil, and Haiti. My weekend explorations had evolved from running images of my face on the AI systems of tech companies to visiting government websites to visually inspect parliament members. I was somewhat surprised at how light-skinned power appeared in Caribbean nations despite their large dark-skinned populations. And I found it interesting to see African countries that I'd been socialized to think of as being behind in the world leading on gender representation. Other explorations I had for potential datasets included looking at Olympic teams. Like the UN, the Olympics brought together a world of nations, yet I also had to attend to the fact that elite athletes were not exactly representative of the typical population in ability, physical form, or age. In some ways an Olympic dataset would bring the same concerns as a celebrity dataset. Parliament members, on the other hand, as representatives of the people, tended to be middle-aged and had a range of body types.

Still, as a former pole vaulter, I couldn't help but continue looking to the world of sports. Teams already so nicely arranged by country, with headshots in often well-lit conditions, on generally well-organized websites, were rather enticing. In looking at professional sports teams and upon visiting the websites of the NBA and WNBA, I was quickly reminded of copyright issues. "Photos cannot be used without the express written consent . . ." The notice regarding consent reminded me of the eerie MITRE dataset that contained images of people who had died in prison. Athletes who signed with professional sports teams gave those organizations the right to use and profit from their images. Those who died while incarcerated could not have consented to this use of their images. I doubt their families even knew their images were being used in this way. Even for the celebrity dataset, many of the subjects contained in the datasets were ignorant of their inclusion.

IBM faced fire for use of a subset of YFCC100M, a dataset released by Yahoo containing 100 million photos under a Creative Commons license on their Flickr platform.[8] Many people did not know their faces were being used in a research database that had been repurposed for IBM's Diversity in Faces (DiF) dataset, which took a subset of about 1 million images from YFCC100M. A person might upload an image for one thing, like a Flickr photo album, and their face might end up in a dataset for something else. IBM was hardly alone, but their dataset represented a poignant example of extremely common practices in the field. I started looking at the copyright rules for the parliaments I had chosen. Some carried a provision that as public material the content of the website could be used for research and educational purposes, but not all sites had such a provision, including Rwanda's and Senegal's. I reached out to the Boston University Technology Law Clinic for help. They looked at the copyright laws for all the countries I targeted and assessed that as long as I did not redistribute the dataset for profit and kept its use for research purposes I should be clear and within the realm of fair use. Nevertheless, lawful use did not overcome the basic fact that I would be using images of people's faces without their consent to do this research, unless I could somehow obtain the consent of the 1,270 individuals who would be included in the dataset. Despite my struggle with this question, other computer vision researchers I spoke to felt these questions were irrelevant. Their position was: The photos are out there, and they are photos of public officials. What is your issue? With some hesitation I proceeded. The lawful yet awful standard allows practices that have questionable ethics. Maybe we needed a higher standard than "What are my chances of getting sued?"

Another gap I found puzzling concerned how images of people's faces were treated. In graduate school I was required to undergo human subjects research training. Generally this kind of research has a direct interface with individuals who would be completing surveys or interviews. Much of the training focused on consent, privacy, and seeking beneficence to the participants. The training addressed notable failures, like the U.S. government's Tuskegee Syphilis Study and Nazi experimentation. For my research into how AI systems analyze human faces, the computer science researchers I spoke to seemed confused as to why I would need human subjects' approval. Though medical images and biometric information were classified as the kind of data needed for human subjects research, photographs of faces in a nonmedical context were not so classified. Even though universities have mechanisms in place for research on human subjects, in doing computer vision research on human faces, which can serve as uniquely identifying biometric information, I was exempt.

I took this exemption and got on with my research, but questions lingered as the deadline for my thesis approached. How could a face be considered anonymous if it inherently contained unique biometric information? Was this another status quo that needed to change? Pushing for this change would make my research much harder, because it was unlikely I could obtain consent for the use of all the faces I had identified for the dataset. But in using the faces, was I any different than Facebook? After all, the company used images uploaded by users without obtaining explicit consent to use them for facial recognition research. With data being such a vital part of AI research and then subsequent products that built on large data stores, the value of data was becoming ever more apparent to me. What might otherwise be inert datasets could be repurposed in a company's effort to train an AI system. The data could also be sold to other companies to support their AI efforts. For example, the Ever app started as a photo-sharing application for families. The company then evolved to offer facial recognition services made possible by the images that had been uploaded by users of the application.[9] This repurposing of data might not have been the intention of the leaders of the company at the beginning, nor were users told that their images would be used in this way. Instead, as with many terms of use, users gave broad nonspecific access to their data, allowing for unknown profitable downstream use. These kinds of loose data practices were targeted by the European Union with the passage of the General Data Protection Regulation (GDPR), which did not extend outside Europe. As I was making the dataset in 2017, none of these protections were in full effect. Data posted on the internet appeared to be free for all computer vision researchers. I was not aiming to profit off the faces I had identified for my research. I also reasoned that as public figures, not private citizens, the subjects had already made a choice to be in the public eye.

Furthermore, nothing prevented me from downloading images and testing them on AI systems for private use. In 2017, I downloaded images of parliamentarians from around the world. I requested common face datasets from other researchers that had been collected through scraping websites without explicit consent. Such a practice was lawful. Why should I feel awful? For the images I ran on gender classification and age models on my own computer, there was some validity to the notion of privacy, since only I had access to the data. I saw the results and could decide what to do with them. But for images that were fed to the remote AI systems through online demos offered by tech companies, I could not know how companies might use that data beyond the demo interfaces they provided.

Even using Creative Commons images that were uploaded to the internet did not sidestep the fact that photographers might post images they had taken without consent. Even if consent had been granted to upload these photos, there certainly was not affirmative consent that years later these images could be used to power data-hungry machine learning models.

In other words, the status quo was built with a disregard for consent.

That disregard for consent is at least in part for the sake of expediency. A researcher cannot quickly collect millions of face images if permission must be requested for every image. A large tech company can create a platform for uploading images, can make it part of the terms and conditions that those images may be used, and can therefore lawfully use those images as desired— still without having explicit consent. Holding these tensions, problematic demographic and phenotypic labels, and unconsented if still public images of public officials, I pressed onward. With the classification in hand and a set of images of parliamentarian faces collected from official government websites, I had one more major step before I could use the data to test any AI systems. The face images needed to be labeled with gender and skin type. Instead of employing Mechanical Turk workers to do this task, I decided that 1,270 images was a small enough set for me to handle, although the work would be tedious. It provided more hard fun for the weekends.

Behind the headlines and slick marketing pages, so many of the advances in machine learning are dependent on labeled datasets. In their book *Ghost Work*, Mary L. Gray and Siddharth Suri talk about these often forgotten workers whose labor is a crucial component of the process of machine learning and who also support research efforts. My process of labeling my new dataset, which I called the Pilot Parliaments Benchmark, gave me just a small taste of that work. And though at first I wanted to skip the tedium and delegate data operations to an undergraduate to do the essential yet undervalued and underpaid work, the process of labeling the dataset myself by hand provided its own insights. When I was crawling through labels to use for my research, I learned that the person deciding the classification systems holds power. Having collected a dataset in need of labeling, I was in position to exercise another form of power—the power to label.

It was unsettling to label the faces of other people. My decisions would impact the ability to robustly assess the target gender classification systems. If my dataset selection and the labels were heavily skewed (like the datasets that came before) or poorly labeled, my experimental setup might be insufficient to surface any indications of bias even if they were present. Like a doctor using the wrong diagnostic tool and thus missing signs of a disease, I would be in danger of missing the mark if my dataset was not well constructed. Yet, like the doctor, I would still have the authority to say, "Look, I ran a test and found no problem." Currently, I was acting as a second opinion on what had been increasingly accepted in the computer vision research community concerning the performance of facial recognition technologies.

To provide that second look, my methods had to offer a different perspective. When we look at proclamations about the performance of AI systems, we must have insight into the types of tests that were used to reach the conclusion, the data collection methods, the labels used, and who was involved in decision-making. Yet as I went through each face in my Pilot Parliaments Benchmark to determine gender and skin type, the subjectivity of the exercise became more apparent. We cannot ignore this subjectivity as more AI systems are introduced into our lives, and we need to push for high standards for AI deployers to support the claims they make about what their systems can do. We also need redlines and guardrails that prevent harmful if unintended consequences from the use of data.

With the rise of generative AI systems that produce images based on text prompts or uploaded example images, we are seeing more mainstream discussions about privacy and data consent. For example, the 2022 release of the Magic Avatars feature in the popular Lensa application increased public interest in AI-generated images. For a fee of four dollars or more, users could upload several images of their faces to the platform and the Magic Avatars system would return fifty or more AI-generated profile images. Social media accounts of celebrities, influencers, and everyday people began to overflow with stylized profile images that appeared to be created by skillful digital artists. Sometimes someone would generate images of another person. For example, Katie Couric's husband, John Molner, uploaded photos of her to the app and then shared the results, which she posted to social media. The trend might have appeared like innocent fun to onlookers and participants. What's the danger in getting stylized profile images for less than the price of a fancy coffee?

Shortly after Magic Avatars gained popularity, some women started noticing that the avatars generated for them included hypersexualized images. Some women received avatars with their likeness depicted on scantily clad bodies or completely topless. Melissa Heikkilä wrote for the *MIT Technology Review,* "My avatars were cartoonishly pornified, while my male colleagues got to be astronauts, explorers, and inventors."[10] When she and other women uploaded their images, I doubt they had envisioned the AI system would use information gathered from their faces to then create renderings of exposed breasts. AI-generated images have opened up another way in which misogyny and the male gaze can be used to deprive women of dignity. Campaigns against digital forms of gender-based violence like revenge porn are increasing. Revenge porn involves the use of intimate images shared with a partner, who then circulates those images to humiliate or bully a person who intended the images to stay private. Generative AI systems make it possible for someone like a jaded lover or stalker to generate sexualized images of you without your knowledge or consent. Deepfakes, AI-generated photorealistic images and videos, have already been used to superimpose the faces of celebrities onto the bodies of individuals performing sexual acts without any regard for consent. The users who spoke up about Magic Avatars, with its more cartoon-like depictions, were not using the system with the intent of generating sexualized content. Olivia Snow, for example, reportedly uploaded childhood photos and received sexualized images of herself as a child.[11] While this outcome was likely not the intent of Prisma Labs, the makers of the Lensa app behind the Magic Avatars feature, it does, beyond the ethical issues, open serious legal questions.

Storing and distributing images and videos of child pornography is a criminal offense in the United States. Who is culpable if an AI system produces sexualized images of an adult when they upload childhood photos? What stops a nefarious actor from intentionally uploading the image of a child to get sexualized images? At the time of writing there are no laws that ban the generation of sexualized images of children from AI systems. Governments and companies have a responsibility to do more. Other AI image generators have features like an NSFW (not safe for work) filter, in the case of DALL-E, a text-to-image-generation system, or community guidelines to not create offensive images, like the Midjourney platform. However, technical solutions like the NSFW filter are not flawless, and community guidelines rely on the goodwill of participants. Both options are insufficient on their own, because ultimately there must be accountability for the creation and propagation of illegal or harmful imagery. If companies were to be fined or have their systems shut down should they produce illegal images, it would disincentivize the release of apps and features that have not been well tested and fine-tuned to prevent harmful depictions.

You might be wondering how these AI images get generated in the first place. The answer goes back to datasets. Generative AI systems are trained on datasets of images gathered from the internet. LAION-5B is an open-source dataset of digital art that includes artwork uploaded by artists in the collection of 5.85 billion images.[12] The Magic Avatars feature relied on a subset of this freely available dataset. Unknowingly, creatives who shared their art to build career opportunities had these images taken and used in systems like Magic Avatars. Images uploaded to the internet for one purpose are often repurposed without explicit consent. Artists who already struggle to make a living based on their creative practice have expressed alarm at the use of their work to fuel generative AI systems. Economic fears around human digital artwork being devalued, and legal questions around ownership and copyrights, have led to websites like Have I Been Trained (haveibeentrained.com/). This "opt-out" website helps artists see if their work has been used in open-source datasets and lets them request that the data be taken down. Similarly, Adam Harvey and Jules LaPlace created Exposing.ai to allow people to see if their faces were included in open-source face datasets and can thereby facilitate deletion requests.

This opt-out approach is a stopgap solution absent regulations and a transformation in the creation of AI systems. Instead of putting the burden on individuals to retroactively see if their images have been used in systems created without their knowledge, opting in should be the default. By the time someone finds out their images have been used, the company has already trained the AI system, so deleting the image doesn't delete its contribution to training the AI system to do a specific task. Even if the image is deleted, the many copies of the dataset already made still contain the images. This is why when Meta (then Facebook) announced the deletion of nearly 1 billion faceprints, there was pushback alongside the celebration. The celebration was around the fact that a major tech company deleting the faceprints was an acknowledgment of the risks associated with face-based technologies that many organizations, including the AJL, had been highlighting. But we do not know if someone at the organization made a secret backup copy, and we may never know. Facebook's actions provided a counternarrative to the assumption that once a company has your data there is nothing that can be done about it. Public pressure makes a difference, and so too does legislation and litigation.

While the company deleted the faceprints, they did not delete the facial recognition models created with those faceprints. The company had legal reasons to rid themselves of the faceprints. In 2021, Facebook reached a $650 million settlement for allegations of violating BIPA, the Biometric Information Privacy Act of Illinois.[13] Crucially, faceprints are derived from images of a person's face, but they are not a copy of the uploaded images. Think of the faceprint as a digital representation of your face. Multiple uploaded images can be used to generate one faceprint. Thus a company can delete the images you uploaded and still keep data about your face in the form of a faceprint. The company can use that face data to train AI systems. Once they have a trained system, they can sell it or use it for research and development purposes.

They can then delete the face data and keep the trained AI system.

Legislation and policies that address user consent and data privacy need to keep downstream uses in mind. We need to think about not just deleting uploaded images and deleting data derived from those images like a faceprint, but also deleting models that were developed with ill-gotten or repurposed data. We need *deep data deletion,* which consists of fully deleting data derived from user uploads and the AI models trained with this data. Commercial AI products should be built only on explicitly consented data sources. We should mandate deep data deletions to prevent the development of AI harms that stem from the collection of unconsented data. Companies like Clearview AI, which scraped billions of photos of people's faces uploaded to public social media platforms, was fined 20 million euros by the Italian Supervisory Authority and ordered to delete the biometrics data of "persons in the Italian territory."[14] The Italian face purge is a start. Deep deletions are possible, but achieving them will require a transformation in the development, auditing, and regulation of AI systems.

Downstream uses of your data should require explicit and affirmative consent, especially when companies like Prisma Labs make money using the repurposed labor of faceless artists. The terms of service of the Lensa app at the time of writing state that the company has rights to the face data of the users who pay for the stylized images. The company claims that data is used only to train their internal AI system. But there is nothing stopping the company from being acquired and that data being used in unforeseen ways. Another consideration is that many online systems are linked to tech giants like Google and Amazon. For the Lensa app, the face data is processed in part using Amazon Web Services

(AWS). AWS powers a significant part of the internet and is part of critical internet infrastructure. Companies like Netflix and Zoom are reliant on AWS.[15] Thus you have to think not just about what one company like Prisma Labs can do with your data, but also what interlinked companies that run the computing systems to process your data can do. There is a daisy chain of companies involved with internet and app-based AI services.

Think about AI systems not as one-off products but as an interconnected ecosystem of datasets, distributed data processing systems offered by big tech companies, and unwitting generators of data that fuel the system: us. Without our uploads the ecosystem would collapse. However, this does not mean it is up to individuals to solve these issues in isolation. I think about major tech companies as infrastructure providers. They currently provide the roadways of internet systems. Because so much of modern life is based on interacting through the internet, and because governments increasingly encourage the use of digital systems, choosing not to use the internet is like attempting to live off-grid. At some point you will likely feel forced to participate, particularly in emergency situations. Within any kind of large-scale infrastructure system there will be breakdowns and maintenance required. Some bridges built using old paradigms that no longer serve the current age need to be torn down completely. Others might be rebuilt, depending on the circumstance. When people point out potholes that can lead to dangerous accidents or show the damage done to their vehicles as a result of a pothole, we don't ask them to stop using roads. We also do not ask individuals to fix the potholes themselves. Instead, we reach out to groups established to safeguard the public interest and to maintain infrastructure. The responsibility of preventing harms from AI lies not with individual users but with the companies that create these systems, the organizations that adopt them, and the elected officials tasked with the public interest. What we can do as individuals is share our stories, document harms, and demand that our dignity be a priority, not an afterthought. We can think twice before participating in AI trends like stylized profile images, and we can support organizations that put pressure on companies and policymakers to prevent AI harms.

CHAPTER 10

ARBITER OF TRUTH

I still had a few more months of experimentation before the end
of my second year at MIT. As I dug deeper into the research, I
was moving further away from what I had assumed would be
a largely technical excavation. Exploring labels around social constructs like race and ethnicity revealed the sociotechnical nature
of the work I was doing. Now that it came to labeling human
faces, I was being confronted with ethical questions about collecting data and philosophical questions about the nature of
truth. To train a machine learning model to respond to a pattern
like the perceived gender of a face, one approach is to provide
labeled training data. The labels represent what machine learning
researchers call "ground truth." Truth as a concept is slippery. For
thousands of years humans have debated what is true, whether in
courts, philosophers' chairs, labs, political rallies, public forums,
the playground, or when looking into mirrors—"Objects are

closer than they appear." Scientists have argued for objective truth that is uncovered through experimentation, yet science does not escape human bias and prejudice. Former scientific "truths" have been debunked: The earth was once accepted in Western science as flat based on observations. Before Copernicus it was accepted that the earth was the center of the universe and all the heavens revolved around it. Women were positioned as intellectually inferior based on observations—men as the center of society, women revolving around them. Once-accepted pseudosciences like phrenology—observing the shape of a head or face to infer internal characteristics—have been shown to serve racial stereotypes.

Feminist scholars have long pointed out how Western ways of knowing, shaped by patriarchy, attempt to erase the standpoint of the observer, taking a godlike, omniscient, and detached view. However, our standpoint, where we are positioned in society, and our cultural and social experiences shape how we share and interpret our observations. Acknowledging that there is subjectivity to perceived truths brings some humility to observations and the notion of partial truths. The elephant can be perceived as many things depending on whether you touch the tail, the leg, or the trunk. This is not to say all interpretations are valid, particularly when looking at physical phenomena. Regardless of your acceptance of physical laws, gravity and the knowledge engineers have gained about aerodynamics influence how we build airplanes. In the world of machine learning, the arbiters of ground truth— what a model is taught to be the correct classification of a certain type of data—are those who decide which labels to apply and those who are tasked with applying those labels to data. Both groups bring their own standpoint and understanding to the process. Both groups are exercising the power to decide. Decision-making power is ultimately what defines ground truth. Human decisions are subjective.

The classification systems I or other machine learning practitioners select, modify, inherit, or expand to label a dataset are a reflection of subjective goals, observations, and understandings of the world. These systems of labeling circumscribe the world of possibilities and experience for a machine learning model, which is also limited by the data available. For example, if you decide to use binary gender labels—male and female—and use them on a dataset that includes only the faces of middle-aged white actors, the system is precluded from learning about intersex, trans, or nonbinary representations and will be less equipped to handle faces that fall outside its initial binary training set. The classification system erases the existence of those groups not included in it. It can also reify the groups so that if the most dominant classification of gender is presented in the binary male and female categorization, over time that binary categorization becomes accepted as "truth." This "truth" ignores rich histories and observations from all over the world regarding gender that acknowledge third-gender individuals or more fluid gender relationships. Two-spirit people have long been recognized in Native American cultures. Hijra or Kinnar people in India can include transgender and intersex individuals.

When it comes to gender classification systems, the gender labels being used make an inference about *gender identity,* how an individual interprets their own gender in the world. A computer vision system cannot observe how someone thinks about their gender, because the system is presented only with image data. It's also true that how someone identifies with gender can change over time. In computer vision that uses machine learning, what machines are being exposed to is *gender presentation,* how an individual performs their gender in the way they dress, style their hair, and more. Presented with examples of images that are labeled to show what is perceived as male and as female, systems are exposed to cultural norms of gender presentation that can be reflected in length of hair, clothing, and accessories.

Some systems use geometric-based approaches, not appearance-based approaches, and have been programmed based on the physical dimensions of a human face. The scientific evidence shows how sex hormones can influence the shape of a face. Testosterone is observed to lead to a broader nose and forehead—but there are other factors that may lead to a particular nose or forehead shape, so what may be true for faces in a dataset of parliamentarians for Iceland does not necessarily apply to a set of actual faces from Senegal. Also, over time the use of geometric approaches for analyzing faces has been shown to be less effective than the appearance-based models that are learned from large labeled datasets. Coding all the rules for when a nose-to-eye-to-mouth ratio might be that of someone perceived as a woman or biologically female is a daunting task, so the machine learning approach has taken over. But this reliance on labeled data introduces its own challenges.

The representation of a concept like gender is constrained by both the classification system that is used and the data that is used to represent different groups within the classification. If we create a dataset to train a system on binary gender classification that includes only the faces of middle-aged white actors, that model is destined to struggle with gendering faces that do not resemble those in the training set. In the world of computer vision, we find that systems trained on adult faces often struggle with the faces of children, which are changing at a rapid pace as they grow and are often absent from face datasets. In many countries there are protections on the use of images of children, making it more likely that training datasets used by academic researchers will include only adult faces. Private companies that have large stores of children's faces from images uploaded by proud parents and doting relatives can nonetheless still collect, maintain, and even use these images to develop machine learning models.

The point remains: For machine learning models data is destiny, because the data provides the model with the representation of the world as curated by the makers of the system. Just as the kinds of labels that are chosen reflect human decisions, the kind of data that is made available is also a reflection of those who have the power to collect and decide which data is used to train a system. The data that is most readily available often is used out of convenience. It is convenient for Facebook to use data made available through user uploads. It is convenient for researchers to scrape the internet for data that is publicly posted. Google and Apple rely on the use of their products to amass extremely valuable datasets, such as voice data that can be collected when a user speaks to the phone to do a search. When ground truth is shaped by convenience sampling, grabbing what is most readily available and applying labels in a subjective manner, it represents the standpoint of the makers of the system, not a standalone objective truth.

I experienced this issue firsthand as I started the task of applying ground truth gender and skin type labels to my freshly collected Pilot Parliaments Benchmark. With the gender labels, I visually inspected the image, looking not just at the face, but the presence of facial hair and the kind of clothing being worn—often suits with the male politicians, but there were some images that did not fall so easily into my visual stereotypes for men and women. There were cases where my visual expectation of gender presentation norms was broken. Here I would examine the text description of the individual available on the government website, looking for titles like Mr., Ms., or Mrs. for clues; if those titles were not available, I would look at the pronouns used when there were biographic descriptions. I was using data beyond the image to try to make my best guess. Guessing at a ground truth is already a sign you are on shaky ground. I stopped saying I was labeling gender and started saying I was labeling perceived gender, which from my experience of doing the labeling was a more apt description. Perceived gender introduced the notion that someone was doing the perceiving, and this perception might not be the so-called truth of the matter.

The experiences reminded me that when I was growing up, I was often asked if I was a boy or a girl, especially in middle school. Other kids would stare and make rude comments. Even adults got confused. Some commended my parents on their two handsome sons when looking at my older brother and me. I remember a substitute teacher, doing roll call, asked if "Joy" was present. In baggy clothes, with my hair braided back in cornrows and a faint mustache beginning to sprout, as is common for many members of the human species despite what beauty standards dictate, I did not look like fellow tween girls. I answered "Here" in a low voice. She looked incredulous and asked again if "Joy" was present. Just as I had trouble labeling faces that didn't fit my visual expectation, she was trying to square the name "Joy" with the apparently boy-clad kid sitting in the back row. I had mixed feelings. On one hand, I was intentionally rejecting fitting gendered expectations about how I "should" look or sound as a "girl child." So in that regard I had succeeded. However, having my name questioned compounded all the teasing I already faced and made me feel self-conscious.

Despite the sting of misclassification, now here I was enacting the binary question: "Is this the face of a man or a woman?" Who gets to define, who gets to label, who has the power to exclude? And who doesn't? In this case, I was the decider. Still, my confidence in my decisions was at times weak.

In addition to applying binary gender labels, I also hand-labeled each face with what I considered to be the appropriate Fitzpatrick skin type on a six-point scale. Because there was already a precedent for applying race and ethnic labels, I felt if I used the familiar labels, I would be grappling with widely discussed topics about racial and ethnic discrimination. Instead, I was applying skin type labels from the Fitzpatrick scale, and I hesitated, because from personal experience I knew how skin color was often used for discrimination. While many talk about racism (discrimination and prejudice based on perceived race), few talk about colorism (discrimination and prejudice based on actual skin color), which can happen within a delineated racial group. Proximity to whiteness or lighter-colored skin comes with privileges even within a subordinated racial group.

As I was growing up fascinated by Hollywood, the Black actresses, in particular, who got desirable roles for the most part would be considered light-skinned. Social clubs and sororities established by elite Black organizations in the United States even used the brown paper bag test as a form of social exclusion. Being too black, darker than the brown paper bag, meant rejection based on skin color.[1] Spending time with immigrant Indian friends, I learned of other forms of colorism around the world, which, as in the United States, influenced who was perceived as a suitable marriage partner, employee, or leader. For people considered multiracial, there is also exclusion based on never quite belonging in any particular group. Knowing the impact of skin color and having particular sensitivity since my own skin is on the darker end of the spectrum, I also experienced the phenomenon of people being offended if they were perceived as darker than they perceived themselves to be.

Too many times I endured the experience of people putting their arm next to mine to confirm to themselves that they were or were not as dark as I am. If I flipped over my wrist to the part of my skin that was least tanned and they had a similar color I could sense the disappointment, because after all I was African. The African American children who tested their skin color against mine this way seemed to be trying to prove objectively they were not as "Black" or as dark-skinned as I was. The implication was clear. To them dark skin like mine was undesirable; thus, being lighter than me would mean they were more desirable. At home my mother filled our house with beautiful dark-skinned people, including photos of my relatives, and told me to ignore the non-sense. It was still painful to know that I was considered less than by some simply because of my skin color. Not to mention that the assumption that because I was from Africa I must have the darkest skin was ignorant of the ethnic and skin diversity across the continent.

That the skin color hierarchy was clear even to children was made quite visible in the doll test. For the doll test, children were shown a deep brown–colored doll and a cream-colored doll and asked to point to the doll that was smart, beautiful, or nice. Then they were asked to point to the doll that was dumb, ugly, or mean. Both Black and white children associated the lighter doll with positive traits and the darker doll with negative traits.[2] This example was used in the landmark Supreme Court decision *Brown v. Board of Education* to support the argument that a society that inculcates inferiority based on race further amplified by segregation can breed self-hatred.[3] The test was repeated with a set of images of dolls that used a Fitzpatrick-like scale. The darkest of the six doll images was the one children most frequently pointed to for the negative traits.

The subject of colorism can be taboo, as some see it as divisive in the push for racial justice. The cruelty of colorism is that it recapitulates social rejection and exclusion based on race into a hierarchy based on skin color. Just as a white scholar might shy away from talking about racism and the ways in which she benefits from systemic racism, not many Black scholars who have investigated race and technology have focused on colorism. I wondered if that lack was because some of the leading Black voices, on the privileged end of colorism, did not see it as a topic worthy of discussion, were uncomfortable addressing their own color privilege, or saw the topic as dirty laundry to be kept out of the spotlight. Would I be breaking some kind of code to focus on the appearance of skin and not race? As a dark-skinned woman, I struggled with that question and finally reached out to Timnit for her advice. She would be considered a lighter-skinned African by some standards. Her lighter skin was a source of teasing by some of her friends who had darker skin than she did. But Timnit saw no issue with labeling by skin type. Despite my hesitation, I moved forward. Yes, this work would be uncomfortable, but perhaps it would help us start cleaning some dirty laundry.

The Fitzpatrick scale was based not on color alone, but on how skin responds to UV radiation from sunlight. This factor meant that skin type, while linked to skin color, had more elements at play, namely different kinds of melanin cells. Melanin comes in three flavors: eumelanin, pheomelanin, and neuromelanin. Eumelanin and pheomelanin affect skin color and hair color, while neuromelanin affects the color of the brain. Descriptions of skin type in the Fitzpatrick scale include these: "Type I skin—burns easily, light colored eyes, green/blue." "Type VI, skin never burns." I was curious where my skin type would fall on the scale. I had the false impression that as someone on the darker side I could not get sunburned, until it happened. It was a hot summer day at the Memphis Zoo, and I had spent hours with nothing to protect my exposed neck. I used to become even darker during the summer, especially as I spent time on the track working to improve my pole vaulting or running for new personal bests in the sprint, hurdles, and middle-distance events. But burning, I thought, was something that happened to white people—until I touched my neck. My finger felt like a stinger. From this zoo experience, "never burns" didn't quite fit my description, so I classified myself as Type V instead of Type VI. Perhaps in the back of my mind I also didn't want to be the last doll associated with the worst traits.

I found I had an especially hard time distinguishing Types I, II, and III. For the most part my selection of Nordic countries helped, as many people fit the blond hair, blue- or green-eyed description of categories I and II. For the African parliaments I had assumed most would fit type V or VI, which was mainly true until I got to the South African data. The vestiges of apartheid were still strong; despite being around 8 percent of the population, lighter-skinned South Africans made up nearly 20 percent of South Africans in the Pilot Parliaments Dataset.[4] The hand-labeling of this dataset made me even more skeptical of the fact that NIST uses national visa photos as a proxy for a face-based algorithm's performance on different ethnicities and races. While it might be clear that parliament seats are a reflection of power and privilege, the U.S. visa dataset would reflect not the distribution of skin types in a country, but the distribution of those privileged enough to be able to afford a visa in the first place.

When tech companies started replicating my research methods, I remember a researcher reaching out and asking about specific parliament members, wanting to know how I had labeled their skin. Outsourcing ground truth can make it easier to justify a selection, but it does not necessarily make the classification more correct, just more in line with a precedent that has already been set, even if that precedent has holes. In the case of members who were in the middle of the spectrum with Type IV skin, I noticed that as more people labeled the dataset, the skin itself was not the deciding factor. Instead, facial features, national origin, and perceived ethnicity also played into whether someone on the borderline of classification would be placed in a lighter or a darker category. If the person appeared as South Asian they were labeled as Type IV; if they had lighter perceived skin but were in an African parliament and not phenotypically white, they were placed in a darker category. Like gender, the cases that defied clear-cut classification were those that exposed our assumptions most clearly. For the face images that went into the Pilot Parliaments Benchmark I used for my MIT thesis, the ground truths were a reflection of my assumptions about classifications.

A major part of my work is to dissect AI systems and show precisely how they can become biased. My early explorations taught me the importance of going beyond technical knowledge, valuing cultural knowledge, and questioning my own assumptions. We cannot assume that just because something is data driven or processed by an algorithm it is immune to bias. Labels and categories we may take for granted need to be interrogated. The more we know about the histories of racial categorization, the more we learn about how a variety of cultures approach gender, the deeper we dive into the development of scientific scales like the Fitzpatrick scale, the easier it is to see the human touch that shapes AI systems. Instead of erasing our fingerprints from the creation of algorithmic systems, exposing them more clearly gives us a better understanding of what can go wrong, for whom, and why. AI reflects both our aspirations and our limitations. Our human limitations provide ample reasons for humility about the capabilities of the AI systems we create. Algorithmic justice necessitates questioning the arbiters of truth, because those with the power to build AI systems do not have a monopoly on truth.

CHAPTER 11

GENDER SHADES

While no one has a monopoly on truth, in the tech sector some companies control a significant portion of the ecosystem. These tech giants have enormous resources and influence major areas of our day-to-day lives. Google's search engine sits as a gateway to the world's online knowledge, amassing valuable search history data while serving up profitable ads. Microsoft, with its Windows operating system, is a staple across many businesses who use the software that comes preloaded on the majority of personal computers. IBM, the big blue giant, led the way in providing enterprise and government technology solutions. Amazon is a tech giant that has learned how to dominate many areas in e-commerce while also providing technical solutions for other companies through Amazon Web Services.

In the land of these giants, I had to navigate carefully. Funding for AI research at leading universities was often sponsored by these companies. It was not lost on me that the Stata Center at MIT, which housed the Computer Science Artificial Intelligence Lab, had a Gates wing named after the founder of Microsoft. Some companies, like Google, IBM, and Microsoft, provided fellowships that would pay for or significantly subsidize the cost of completing graduate programs in computer science. After finishing a computer science degree, top students found competitive and compelling job offers from the tech giants, who were ensured a steady supply of promising research talent. Between financing specific areas of research, providing the means for earning expensive graduate degrees, and offering lucrative tech sector jobs, the shadow of the tech giants touched all aspects of the AI research ecosystem.

Getting on the bad side of one of these companies could have serious career consequences. When I was a graduate student, however, I was not particularly interested in working for one of them. I had my tuition fully covered and received a stipend. The funding structure of the Media Lab, and particularly my affiliation with the Center for Civic Media, gave me more freedom than most graduate students researching AI systems would have had. My advisor, Ethan, was also not beholden to any of these tech companies to directly fund our research. Instead, Ethan sought support from organizations like the Ford Foundation, the Knight Foundation, and the Robert Wood Johnson Foundation. All this to say, when it came time for me to select target systems to examine for algorithmic bias, I did not feel limited in whom I could select. I was eager to see if a more diverse dataset could expose cracks others had missed.

With my labeled Pilot Parliaments Benchmark in hand, it was time to pick my targets. For the commercial systems, I focused on companies that had public demonstrations available on their websites that included explicit gender classification. This focus led to the selection of IBM, Microsoft, and Face++. At the time, on the IBM website you could upload images to its Watson project, and on the Microsoft site you could upload images on the Azure demo website. I included the Face++ system from the Chinese company Megvii because of China's major role in AI development. Prior studies had shown that systems developed in China differed in performance on Western and Asian faces, so I wanted to see if there might be any differences between test results from U.S.-based companies and a company based in China. The company had an online demo that was open to the public and also, according to the marketing materials, was being used by many developers.

While these companies all had public demos that I used as an indication of their confidence in their product, uploading 1,270 images by hand would not be practical. To conduct the study I signed up for accounts so I could access their systems with code I would write. The code uploaded each image of the Pilot Parliaments Benchmark to be processed remotely by IBM, Microsoft, and Face++. Once each image was processed I would receive the results, which would include some indication of the face detected and guesses about age, gender, and other attributes depending on what was offered by the particular service. My decision to focus solely on gender was in part to keep the study as straightforward as possible.

The Pilot Parliaments Benchmark I created at MIT was small but mighty enough to show significant oversights and gaps in the evaluation of machine learning models. For comparison, other influential benchmarks usually had tens of thousands of images on the low end and as many as millions of images on the high end. Just because you can doesn't mean you must. Less can be more.

Once I ran the faces against the gender classifiers of the targets, the analysis of the results began. Prior studies looked at overall accuracy. This high-level aggregate view can mask important performance differences between groups. For the published 2018 "Gender Shades" paper, the overall performance did not seem to reveal too much reason for concern. For the Microsoft product, the system had 93.7 percent overall accuracy on the dataset. For Face++ overall accuracy was 90 percent, and for IBM the performance was 87.9 percent. My hunch was that looking at overall accuracy was not enough, even though doing so was the norm. I suspected that much of the research that had been published using only aggregate data gave us a false sense of the progress being made in the development of facial recognition technologies.

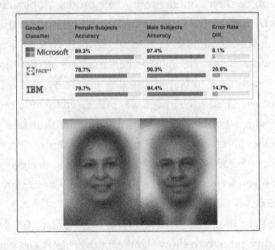

Gender Classifier	Female Subjects Accuracy	Male Subjects Accuracy	Error Rate Diff.
Microsoft	89.3%	97.4%	8.1%
FACE++	78.7%	99.3%	20.6%
IBM	79.7%	94.4%	14.7%

The next stage of analysis focused on comparing performance between male-labeled faces and female-labeled faces. All companies overall performed better on male faces than female faces. Microsoft had the smallest accuracy gap with an 8.1 percent difference. Face++ had the largest gap with a difference of 20.6 percent.

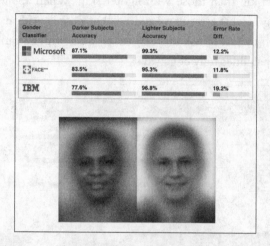

Gender Classifier	Darker Subjects Accuracy	Lighter Subjects Accuracy	Error Rate Diff.
Microsoft	87.1%	99.3%	12.2%
FACE++	83.5%	95.3%	11.8%
IBM	77.6%	96.8%	19.2%

I also examined the performance by skin type. The focus on skin type was chosen not to avoid talking about race, but to emphasize that race as a social construct was too subjective to use as a proxy for the phenotypic characteristics of an individual. Though the dataset was labeled by six skin types, to account for off-by-one errors, I grouped skin types I–III, which covered mainly the parliament members from Iceland, Finland, and Sweden, as the lighter-skinned group. I categorized skin types IV–VI as the darker-skinned group, which covered many of the parliament members in Senegal, Rwanda, and South Africa. Taking this step of phenotypic analysis at the time was a significant contribution to the field of computer vision. Phenotypic analysis demonstrated not just additional ways of understanding gender classifiers like the ones I was testing from these commercial products, but also how we might understand other areas of computer vision. For example, subsequent studies showed that pedestrian tracking systems were more likely to miss detecting people with darker skin than those with lighter skin. When applied to self-driving cars, the result would be that people with darker skin were at increased risk of being hit by a car with automated driving capabilities fully engaged.

Gender Classifier	Darker Male	Darker Female	Lighter Male	Lighter Female	Largest Gap
Microsoft	94.0%	79.2%	100%	98.3%	20.8%
FACE++	99.3%	65.5%	99.2%	94.0%	33.8%
IBM	88.0%	65.3%	99.7%	92.9%	34.4%

The results of the phenotypic analysis showed that overall, Microsoft, IBM, and Face++ did better on lighter-skinned faces than darker-skinned faces when it came to guessing the gender of a face. If we were to stop at this level of single-axis analysis—that is, looking at gender in isolation and skin type in isolation—the assumption would be made that when it came to gender, regardless of skin type, systems performed better on men than women. Conversely, without digging deeper we might also assume that when it came to skin type, performance would be better on lighter skin than darker skin regardless of gender. These assumptions, which you might attempt to justify with the single-axis analysis, would, however, miss a more complex story. We needed to go another step further in the analysis, because we cannot assume, if the data has not been collected and tested, that the system performs the same across all groups. Furthermore, the burden of proof of performance needs to be placed on the people developing systems, not those who are impacted by their use.

To go beyond single-axis analysis, I broke the Pilot Parliaments Benchmark into four intersectional groups: darker-skinned females, darker-skinned males, lighter-skinned females, and lighter-skinned males. While it was true in this study that all companies did better on male faces than female faces overall, and they did better on lighter faces than darker faces overall, this knowledge was not enough to know how the different companies would perform at the intersections.

For Microsoft, the intersectional breakdown revealed that perfect performance was achievable for one group, the pale males: 100 percent accuracy. The company performed worst on darker-skinned females: 79.2 percent accuracy. Darker-skinned males had 94 percent accuracy, and lighter-skinned females had 98.3 percent accurate performance.

For Face++, the intersectional method showed that the best performance in this case was on darker males at 99.3 percent, followed by lighter males at 99.2 percent. The worst performance remained on darker females at 65.5 percent. For IBM, the best performance was on lighter males, followed by lighter females and then darker males. Like its peers, IBM had the worst performance on darker females: 65.3 percent.

The intersectional analysis showed that the largest gap in accuracy was up to 34.4 percent between lighter males and darker females with the IBM classifier. Such a disparity in performance was not captured in the aggregate performance of IBM at 87.9 percent accuracy on the entire benchmark. Nor would this kind of gap be apparent in the gender gap of 14.7 percent or the skin type gap of 9.2 percent for IBM. The different patterns show the importance of examining individual systems, since we cannot make assumptions about which way intersectional analysis will fall.

What accuracy doesn't reveal are questions around failure. When a system fails, how are the errors distributed? We should not assume equal distribution of errors. Beyond just looking at the accuracies of a system, we can also learn more about performance by looking at the kinds of errors that are made. For error analysis, I took it one step further and looked at the results by each skin type, revealing even worse performance. When I looked at women with the darkest skin on the Fitzpatrick skin type scale, I found that for female-labeled faces of type VI, error rates were up to 46.8 percent.

To my knowledge at the time these were some of the biggest accuracy disparities recorded regarding commercially sold AI products. My test had exposed significant cracks in the system not just from small research labs but from some of the oldest and most respected tech giants. If the tech giants had bias in gender classification, what other biases could be lurking in their AI products?

AFTER TWO YEARS ON THE MIT CAMPUS, I WAS FINALLY MAKING my way to the Hayden Library. With online access to almost all the resources I needed, I had yet to step foot into a campus library. After printing out results on archival paper that felt soft yet sturdy to the touch and hunting down last-minute typos, I felt ready to submit my master's thesis. During the summer months of 2017, I went back and forth over datasets, spreadsheets, and calculations. I eagerly emailed my results to my committee members and was surprised by the comments they left in the margins.

I paid a visit to Hal Abelson. "Your calculations do show differences in accuracy, but what's the harm? Why should people care?" Hal's questions lingered with me. "Is being misclassified harmful?" "Does it matter if a parliament member is misgendered?" Maybe the example I had chosen was not compelling enough, or maybe having a committee of all men made them less sensitive to the implications of misgendering women?

To me this was akin to announcing, "She was hit by a brick; look at the evidence" and then being asked, "How do we know being hit by a brick is harmful?"

I could not assume everyone would see that the issue with gender classification was problematic in and of itself and also indicative that there could be similar issues in other areas of facial recognition technologies and AI. Being misclassified as a criminal suspect and being wrongfully arrested certainly seemed high stakes to me. The common thread was that the deep learning techniques being used to create gender classifiers were also being used to create systems to attempt to identify a unique individual or detect skin cancer. All these systems needed to be assessed for algorithmic bias. There were also harms of dignity due to the misclassification of someone's gender, but those harms didn't seem to land with the members of my committee.

I was frustrated because results that I thought were substantial now felt diminished. Maybe I was making too big a deal of what I had uncovered. Or maybe this was part of the academic rite of passage that required you to defend the importance of your work to a skeptical audience of senior academics.

I lobbed back questions of my own to Hal and the rest of the committee. "Should I show the faces of those who are mislabeled?" None of the committee saw an issue, but showing the mislabeled faces directly didn't sit well with me. It felt like I would be framing an insult that had been applied to someone. I decided to show average faces, composite images that blended multiple people together so no specific identity could be seen when I presented specific numbers on errors. I also asked my committee, "Should the thesis focus be on algorithmic bias or dataset bias?" In other words, was the work more about data or more about the algorithms involved in the creation of the gender classifiers I tested? Over time the answer was both. The dataset skews in benchmarks led to a false sense of progress, and existing popular datasets had biases that needed to be explicitly named. The gender classification systems that were based on machine learning models were influenced by the dataset, yes, but preprocessing steps like face detection could introduce bias into the process as well. Subsequent research showed that for some systems, even when datasets were more balanced in the training portion, outcomes would be skewed, indicating that data by itself was not the only piece of the puzzle.

Over the summer, Hal was teaching me that I could not assume everybody would interpret gender classification or misclassification as being harmful. He was pushing me to provide more context and articulate why the work mattered. He was also preparing me for the skepticism that others in the computer science community would have. At the same time, I was not doing the work for the computer science community alone, but for those who could find themselves on the wrong side of a label.

Hal guided me to emphasize that the techniques used for gender classification were being used in other areas of computer vision and machine learning, and therefore these findings were a warning sign. "This might seem obvious to you, but you are going to need to draw the connections so people understand what you are talking about." Though the results were focused on gender classifications, the implications had other repercussions. Still, I did not want to minimize the impact of misclassification. I thought back to all the embarrassing times in my childhood when I was asked if I was a boy or a girl, or I was assumed to be a boy. I also did not want misclassification to be taken as a call to simply create better gender classifiers, but rather to look more closely at the enterprise of labeling people in the first place. These considerations would be ongoing conversations. For now, I had a master's thesis to submit.*

On my way to Hayden Library on the MIT campus, I saw a familiar figure, Dr. Ruchir Puri, the chief scientist at IBM Research. We had met at the second Aspen Institute Roundtable on Artificial Intelligence earlier in the summer, where I had briefly shared my research focus with him and enjoyed our conversation about the future of AI. I began to speed walk over the red bricks that accented familiar sidewalks. Holding up my bound papers, I shouted, "Ruchir! I am submitting my thesis. You are going to want to see the IBM results." He stopped and turned with his two daughters. He introduced them to me with fatherly pride, and then said with anticipation. "Send me a copy!"

I would *eventually* send Ruchir the results. By the time I submitted my master's thesis, I had decided to continue on the aca-

---

* My 2017 master's thesis not only audited AI systems from big tech companies but also introduced the algorithmic inclusion scorecard (p. 99) and the dataset diversity scorecard (p. 100). These concepts were precursors for later papers that introduced "Datasheets for Datasets" to characterize dataset diversity and "Model Cards for Models" to characterize performance on machine learning models across diverse populations. www.media.mit.edu/publications/full-gender-shades-thesis-17/.

demic path and had been admitted into the MIT Media Lab PhD program with Ethan as my advisor. However, Ethan was taking a year off to work on a new book, and so I would be spending a full year with the Lifelong Kindergarten group under the welcoming eye of Mitch Resnick. With my master's thesis finished, I could use the first semester of the PhD program to work on publishing the "Gender Shades" results.

The timing could not have been better. That summer, the Conference on Fairness, Accountability, and Transparency in AI had been announced. They were looking for submissions for the inaugural conference that would take place in February 2018. Timnit and I collaborated to submit the "Gender Shades" paper to the conference. While some reviewers questioned the novelty of the findings and the need to use the Fitzpatrick classification systems, the paper was eventually accepted. A peer-reviewed publication was a big accomplishment for me as a graduate student. It added additional credibility to my thesis work. I also hoped it would make the tech companies take me seriously when I reached out to them.

Now with an accepted paper, I began sharing the results with all the companies. I emailed IBM's Ruchir as well as Microsoft and Face++ representatives, since their companies were all implicated by the paper. I heard back from Ruchir almost immediately. He wanted to know who the competitors were. I had sent the paper results with all the findings, but instead of including the company names I used companies A, B, and C. They would have to wait until the conference to know for sure who their competitors were. In the meantime, IBM took a very proactive stance. Since I was not distributing the Pilot Parliaments Benchmark dataset, they created an internal dataset of their own and shared with me that they had observed similar findings. They invited me to their New York office on January 26, 2018, to speak to a number of executives.

A few days after turning twenty-eight, I walked toward the IBM Watson building in New York City. I was greeted by a large metallic red balloon dog. The Jeff Koons statue welcomed me to corporate America. After passing the security guards and rows of cubicles, I entered a clearing that looked like a small amphitheater, where I would walk through my research findings. I presented the "Gender Shades" work to Ruchir Puri, IBM chief scientist; Francesca Rossi, head of AI and Ethics; Anna Sekaran, head of communications; and members of the computer vision team, who sat nervously as I advanced through the slides showing performance results.

"Who are the other companies?" they queried.

"You are going to have to wait and see . . ."

After the formal presentation, I enjoyed a lunch with the team. Again I was asked, "Who are the other companies?"

The meal was good, but it wasn't *that* good. "You will have to wait and see."

Later that evening I went out for a truly delicious Italian meal with a few IBMers who had been part of the drill team examining my research. For a third time, I was asked, "Who are the other companies?" As I took another spoonful of pasta, I nearly slipped—but then I caught myself.

"You are going to have to wait and see . . ."

I savored teasing the IBMers about the unnamed competitors and also having a chance to talk shop with the computer vision team. Clearly, I was not there to try to line up a job at IBM after graduating. One member shared an insight: "You know, when we replicated your approach and ran it on our old model, our results were even worse than what you reported." I suppressed a smile. I had intentionally designed the study such that when others repli-cated it, they would likely get worse results. I did not publish the worst possible findings in the academic paper.

After the NYC meeting, I had a few meetings in the IBM Bos-ton office. The team let me know that they had developed a new model and wanted me to share the results at the conference. I insisted that I would have to test their new model at my office, so they took the trip to the Media Lab and met me right around the corner from the Foodcam. Their dress clothes gave corporate vibes as they walked past the ping-pong table in the atrium. Sit-ting next to my LEGOS, surrounded by grad school artifacts and the white mask, the IBM team members shared their model with me. The numbers did in fact look significantly better than the older model.

One of the computer vision team members put his hands to-gether in a prayer pose and looked me dead in the eyes. "Please share the updated results." I understood the look in his eyes. His reputation was on the line. Just as there are people whose lives are impacted by the labels and classifications that machine learning systems put out, there are people on the other side of the equa-tion whose careers and livelihoods are linked to the products I was scrutinizing. Before my visits to the IBM offices, the develop-ers of these commercial systems were as faceless to me as the users who are impacted by these systems might have been to them. They were in some ways "just" doing their job in the ways they had been trained to do. But when just doing your job can mean someone else's life chances are at stake, there has to be a higher standard. Doing your job doesn't excuse harm. Looking at me with pleading eyes, he was saying we can do better. He re-minded me of one of the postdocs my father had trained who came to our home and helped set up the Silicon Graphics machines that had fascinated me as a child. Most of my dad's students were immigrants like us. As the daughter of an immigrant researcher, I imagined if my dad were in that position. What if it were my father working for a corporation? What if his livelihood and ability to support our family—as well as his reputation in a field he had devoted decades to—were on the line? Would I share his updated results? The engineering side of me could also sympathize with the challenges of building functional systems. I had only a few weeks to digest these thoughts and make a decision.

CHAPTER 12

DESERTED DESSERTS

As I pushed my book bag under the seat in front of me, I relaxed into my chair, ready to travel from Boston to New York City for the second time in two months. I was on my way to give the first public academic paper presentation of "Gender Shades" at the inaugural FaccT Conference, a computer science conference that brings together researchers and engineers interested in fairness, accountability, and transparency in tech.[1] The anticipation was building inside me. I rearranged my slides multiple times and practiced what I would say under my breath, hoping not to disturb the person sitting next to me. Hoop by hoop I was clearing the academic hurdles needed to one day become Dr. Buolamwini: Master's thesis—check. PhD program acceptance—check. Published academic paper—check. Conference paper presentation—pending.

The conference was held at New York University. After grabbing a registration badge at the front desk, I walked into a reception area. A few tables were stacked with bran muffins, grapes, cheese cubes, and other standard conference snacks. After selecting some food to nibble on, I began to spot researchers whose work I had read about. My conversations with Cathy O'Neil and Timnit let me know I was not alone, and at this conference I hoped to find more allies in this emerging area of study. The reception hall was full of researchers, sponsor representatives, and the occasional journalist signaling the growing interest in algorithmic bias work.

The keynote presenter was Dr. Latanya Sweeney. Her work laid down the intellectual foundation for mine. In 2013 she had published a highly influential paper that demonstrated racial bias in search engine results, called "Discrimination in Online Ad Delivery."[2] The paper showed that search results for names that were more likely to be given to Black babies, compared to names that were more likely to be given to white babies, had a higher chance of having an arrest record advertisement displayed. Though the ad suggested that the person with the name had been apprehended by police, that did not actually have to be true. The consequences of this disturbing pattern aren't hard to imagine: If a landlord or a hiring manager searched for an applicant's name, and then noticed a link to an arrest record advertisement, consciously or perhaps even unconsciously they might decide to pass on the applicant just to be safe. Such ads could play into confirmation bias, given racist stereotypes around Black Americans and criminality. The paper pointed to what are known as both allocative harms and representational harms associated with stigma. Allocative harms refer to the denial of tangible resources or opportunities like a job or housing. Representational harms deal with the stories and images that are circulated about who and what is bad in society. During her conference remarks that day, Sweeney graciously alerted the audience that they should learn more about the work of a new up-and-coming researcher. She then gestured over to me, and I smiled awkwardly as heads turned my way. Her support was welcome, and I also felt increased pressure to live up to her praise.

The night before my presentation, Timnit and I had dinner at a nearby bistro.

"Should we present this together?" I asked her.

"Nah, you are the lead author. Plus, the paper is based on your master's thesis. I will be cheering you on in the crowd. On a different note, are you going to that dinner?"

I had received an email invitation to attend dinner with one of the conference sponsors, but I was still on the fence. Cathy O'Neil lived in New York and her bluegrass band was playing a gig. Maybe I could do both. I was staying at a cramped Airbnb with a fellow Media Lab student and was looking for any reason to spend as little time as possible in the small apartment.

"I'm thinking about it. Are you going?"

"Only if you come with me."

THE NEXT MORNING, I PULLED OUT MY BLACK-AND-WHITE AJL shield. To mark the special occasion, I donned my reflective transparent yellow eyeglass frames and wore yellow feather earrings. I looked less like a computer vision researcher and more like a canary among a sea of black-, navy-, and gray-clad academics. As I walked to the podium my chest tightened. I pushed through the feeling and started the presentation, beginning with my story of the white mask demo. I described the limitations of existing gold standards and how power shadows plague the field. Then I got to the research findings, moving from overall accuracy to accuracy by gender, then skin type, and finally intersectional accuracy. Once I got to the results with the IBM slides, I looked up from my laptop and scanned the audience. IBM's Francesca Rossi was sitting near the front row with a look of anticipation in her eyes.

"Of all the companies I reached out to, IBM was the most responsive. They replicated the paper and actually released a new model a few weeks ago."

I advanced the slides on the clicker.

"Here are their results."

"Previously IBM performed at 65.3 percent accuracy on darker females; the new model according to their internal tests performs at 96.5 percent.* Change is possible."

I had decided that since IBM made the effort to engage with the research, it was right for me to share their updated results as they had requested. Maybe the company could use their influence to push other companies to do internal algorithmic audits. Perhaps there was room for private/public partnership, after all.

I finished the presentation, and the tightness in my chest finally released. I could hear the audience's applause as my eyes met Timnit's. Her smile illuminated the entire front row.

That evening, elevated by the excitement around the "Gender

* AJL's subsequent study on the original Pilot Parliaments Benchmark dataset recorded IBM performing at 88.5 percent. The difference is based on the setting of the threshold cutoff. When an AI system is trained to classify a face, the system can be configured to give a confidence score regarding the label produced. A threshold number can be set to determine when the classification should be accepted. So if a system says it is .8 confident a face image depicts a woman and the threshold is set to .7, the classification would be accepted because .8 is greater than .7. If the threshold was set to .9, the classification would be rejected because .8 is less than .9. Changing the threshold can change the recorded accuracy of a system. Depending on the product, some companies share a confidence score and others do not. Because the other companies I tested did not share confidence numbers, the AJL study used the label that was provided, "male" or "female," without regard for the confidence score. In their self-reported study IBM used a .99 threshold.

Shades" paper, I decided to go to the corporate dinner with Timnit. On arrival, we were escorted to a private room with lush, velvet-lined walls, reserved for special events. We were free to seat ourselves, so I took a center seat right across from the host.

The host cleared his throat to begin the dinner and acknowledge notable guests. "Everyone gathered here is part of a battle to determine the direction of what AI will be. We hold power."

When it was my turn to speak I challenged the host: "Not everyone here has the same power, certainly not the same power as you." Timnit nodded in agreement. The senior scholars around the table seeking funding from the corporate sponsor shifted their eyes nervously.

The host, seemingly undeterred by my remarks, opened conversation to invite discussion about the future of AI. I felt far from my Media Lab office, where I daydreamed about being among the decision-makers. Now I had a seat at the table, and I was not going to hold back. A guest sitting at the very end of the table eagerly proclaimed, "We must harness AI for good! Think about the people in Africa. . . . I have been working there for some time, and we need to make sure we are thinking about AI from their perspective." Right on cue, I heard a familiar voice at the other end of the table. "I do think of Africa. I'm an Ethiopian refugee. Too many times, we have so-called experts parachuting in ideas. The local people have important knowledge and *actually* know what is going on. They should lead the work, not white saviors!" Moving our heads back and forth like observers at a tennis match, dinner guests were getting much more than a decadent meal.

Polite conversation had died. In the middle of the back-and-forth, the person seated next to me had been nervously sawing away at a piece of steak. They finally made their way through the meat only to have the final cut catapult toward the host.

Thud!

Our host did not look amused, but he held his composure a tad longer. Meg Mitchell,* a red-haired and fire-tongued prolific AI researcher, joined in the heated conversation, questioning whether the sponsor's plan to work with health data might lead to some unsavory ethical complications. The host had had his fill. After thanking everyone for coming, he declared that he needed to leave. Other dinner guests also found reasons to retreat.

I glanced over at the printed menu next to my fork and copper cup. One course remained undelivered. I decided to stay, as did Timnit, Meg, and a few perpetually hungry graduate students. The servers walked into the room, balancing plates with enticing treats. They carefully placed a dessert plate at every seat, including the freshly abandoned ones. Once the servers left, we had our choice of deserted desserts: delectable chocolate mousse, cheesecakes, and more. I had plenty to chew on.

AS WE MADE OUR WAY THROUGH THE DESSERTS, I THOUGHT about my own role in parachuting in ideas. Well-intentioned companies and academics with immense influence were putting AI into the world with the aim of doing good. An enthusiastic guest at dinner had implored us: "You are some of the world's

---

* Margaret Mitchell, who also uses the name Shmargaret Shmitchell, and Timnit Gebru would later coauthor the influential paper "On the Dangers of Stochastic Parrots: Can Language Models Be Too Big?" alongside Emily Bender and Angelina McMillan-Major. At the time of the dinner, I did not know the pivotal role a number of the guests would play in uncovering the dangers of AI.

smartest people, making some of the most powerful technology known to mankind. You should use it for good. You should tackle the hardest problems. This is an immense opportunity." The impulse to use tech for good was a familiar moral imperative. I'd crafted a life's mission when I was nineteen years old to do good in the world and maximize my individual potential: to *show compassion through computation*. As an undergraduate, I had volunteered with the Carter Center to work on neglected tropical diseases, with a focus on trachoma. Trachoma is a preventable disease that once led to blindness in places like the United States. It had since been eradicated in many places in the world, but there were still a few stubborn pockets in Ethiopia and elsewhere.

The trachoma project was attractive to me because the problem was tractable. There was a known intervention to prevent the disease: a drug called Zithromax. However, the medication had to be distributed in rural areas that were sometimes hard to reach, and there was always the question of funding. The Carter Center made a deal with Pfizer and partnered with Ethiopian officials with the goal of completely eradicating trachoma. As I walked through the gardens near the Jimmy Carter Presidential Library in Atlanta, Georgia, I saw a piece of art that offered hope—a bronze statue of a child leading a blind elder using the stick they both held. *Sightless Among Miracles* had been donated to the center to underscore the effort to support the global control of onchocerciasis (river blindness).[3] There was reason to be hopeful: Through concerted efforts, the Carter Center team was on its way to eradicating river blindness in a number of countries.[4]

Next on the near-term horizon for the Carter Center was trachoma. To track the effectiveness of the campaign, the center had protocols to monitor and evaluate their efforts. This is where I fit into the story—my early foray into "saving the world." Their evaluations were done using paper surveys that were later digitized via transcription. Sure, one might say I was merely creating digital surveys if you looked squarely at my day-to-day activities. To me, I was fighting the world's fight and *helping eradicate neglected tropical diseases by using novel mobile data collection approaches*. At least this is what I would say in a subsequent application for a Rhodes Scholarship. *You are contributing to a major global health initiative*. This is what I said to myself when the work got tedious.

I traveled to Ethiopia during the summer of my junior year to pilot the MALTRA (malaria and trachoma) mobile surveying application. I developed the application to help bring efficiency to the error-prone process of collecting paper-based surveys. After multiple flights and interminable hours in a Carter Center four-wheel drive traveling past shepherds, some with rifles and others with staffs, and groups clad in white fabrics that framed their beautiful faces, we arrived in Kombolcha, a town in the Amhara region of Ethiopia. It was time to pilot the system. While our hosts assured us that there would be internet access, the speed of the connection was slower than anticipated.

Coding under a mosquito net in the evening, I reconfigured the system to save data locally. I would not be able to upload the data directly to an online database as I had anticipated. I'd built the system in my playroom earlier that summer in my suburban childhood home in Cordova, Tennessee. But the assumptions I'd made in Cordova did not quite hold up in Kombolcha. At the time, the Google Android tablets I was programming did not have Amharic keyboards, yet most of the people who were being surveyed and the health workers needed to use Amharic. Our project team developed a custom Amharic keyboard for the Android tablets that had to be loaded onto every new device. *De-*

faults are not neutral. Years later I met some Android team members and asked why keyboards were not available in Amharic then. The answer was business economics. To Google, Ethiopia was not a priority market.

In the age of AI, who will decide what our priorities are? What assumptions about using AI for good are being made by people far removed from the day-to-day realities of those they aim to help? My desire to show compassion through computation had to be measured by my ignorance of what would truly be of help. I had taken a parachute approach, jumping into a location I knew little about. Though I like to think that I made a meaningful difference by improving the Carter Center's data collection efforts, my experience in Ethiopia revealed the limits of good intentions and the need for local context.

The Ethiopia trip had imprinted on me that the technology by itself was not enough. Instead of asking questions about whether we should use a resistive screen or capacitive touch screen mobile tablet, I started wondering why I had come to Ethiopia to do this project instead of having locals do the technical development work. A few years later, I did a Fulbright fellowship that centered on equipping Zambian youth to make meaningful mobile applications as an evolution in my ongoing mission to show compassion through computation. In 2013, the sub-Saharan Fulbrighters gathered in Addis Ababa, trading stories and listening to program officials tell us that when we returned home we would be viewed as experts on our host countries. One of the Ethiopian hosts, however, reminded us that after spending just seven or so months in a country, we could not credibly think of ourselves as experts and needed to come in with a mindset of asking what the people who live here are already doing that is working. Bit by bit, deficits in my thinking were being uncovered.

Later I encountered the effective altruism movement, which was gaining momentum around the time I started my Rhodes Scholarship at the University of Oxford. Proponents of effective altruism implored soon-to-be graduates not just to donate to causes that might make them feel good, but to do the research and donate to areas that would have the most impact. One path, for example, might be the "earn to give" route, pursuing a lucrative career in investment banking in order to have more money to donate to charities deemed effective by trusted partners. The framing was at first enticing. What is wrong with doing research to maximize the impact of an individual's disposable income for those who were privileged to do so? Cause triage, or absolute prioritization, required a utilitarian view of doing the most good for the most people without fully integrating the question of injustice. Good is also tenuously defined, sufficiently vague to show positive intent but defined in a quantitative manner that values scale over intimacy, while showing little regard for what cannot be measured or counted.

My problem with effective altruism is that the approach entrenches the status quo. Supporting exclusive charities sidesteps addressing the issues that led to the rise of charities in the first place and does not require changing existing power relations or company practices. The movement evolved from pushing for bed nets like the ones distributed as part of the MALTRA program to thinking about threats and harms to future humans. The reasoning went something like this: In the future it is plausible that there could be trillions of humans, and we have an obligation to safeguard those humans as best we can.

According to this viewpoint, known as "longtermism," with adherents known as longtermists, we have an imperative to be good ancestors, to think through what we owe the future and act accordingly. This view collides directly with the advancement of artificial intelligence. Sure, there could be near-term harms from algorithmic bias like what was uncovered with the "Gender Shades" paper, but an even greater problem for longtermists is looking to the future and thinking about existential threats AI poses to hypothetical people who do not yet exist. In other words, longtermists are concerned with the future risk that AI systems might outsmart the humans in charge of economic and political systems and have adverse impacts on billions of people. This rise of the machines could be the fall of man, and thus it represents an existential threat we must prepare for now, or so the reasoning goes. I wonder if the threat is really that more people are going to be harmed or if those with power now fear becoming marginalized by advanced technology. This rise of machine overlords would replace the human overlords whose current decisions already adversely impact billions of people. Longtermists follow a tradition of showing concern for future descendants. Many ancient cultures have emphasized the importance of taking care of planet Earth in order that future generations can breathe clean air and drink from the bounty of nature. However, safeguarding future generations means taking care of present and addressable dangers.

Longtermist thinking isn't isolated to late-night ruminations by eccentrics. For example, Oxford philosopher Nick Bostrom posed the paper clip thought experiment as a way of illustrating why he believes in the need to plan ways to safeguard against super intelligence that emerges from machines. The thought experiment goes as follows: When humans give an AI system a goal to reach, we do not have full control over how that AI system will reach that goal. Bostrom argues that if the goal is to produce as many paper clips as possible, the pathway taken to do so by the AI poses risk. A sufficiently advanced AI agent could use its intelligence to coerce powerful individuals to divert resources and shape priorities to maximize the production of paper clips. Paper clip production, like computer vision systems mistaking blueberry cupcakes for chihuahuas, might seem like an inconsequential, trivial, or cute example at best. But these ideas and examples are presented in elite institutions, used in college curricula, and shared in a manner that shapes the discourse about the future of AI by those who are being groomed to hold powerful positions in companies, governments, and academia. Concerns about future harms of AI that are based on the rise of artificial general intelligence (AGI) with intelligence superior to humans have given momentum to an area known as AI safety. Anthropic, an AI safety company that emerged in 2021, received more than $700 million in funding in less than eighteen months.[5] What might it look like if similar resources were dedicated to existing AI harms that are neither hypothetical nor distant?

The term "x-risk" is used as a shorthand for the hypothetical existential risk posed by AI. While my research supports why AI systems should not be integrated into weapons systems because of the lethal dangers, this isn't because I believe AI systems by themselves pose an existential risk as superintelligent agents. AI systems falsely classifying individuals as criminal suspects, robots being used for policing, and self-driving cars with faulty pedestrian tracking systems can already put your life in danger. Sadly, we do not need AI systems to have superintelligence for them to have fatal outcomes on individual lives. Existing AI systems with demonstrated harms are more dangerous than hypothetical "sentient" AI systems because they are real. One problem with minimizing existing AI harms by saying hypothetical existential harms are more important is that it shifts the flow of valuable resources and legislative attention. Companies that claim to fear existential risk from AI could show a genuine commitment to safeguarding humanity by not releasing the AI tools they claim could end humanity. I am not opposed to preventing the creation of fatal AI systems. Governments concerned with lethal use of AI systems can adopt the protections long championed by the Campaign to Stop Killer Robots to ban lethal autonomous systems and digital dehumanization.* The campaign addresses potentially fatal uses of AI without making the hyperbolic jump that we are on a path to creating sentient systems that will destroy all humankind.

Though it is tempting to view physical violence as the ultimate harm, doing so makes it easy to forget pernicious ways our societies perpetuate *structural violence*. Johan Galtung coined this term to describe how institutions and social structures prevent people from meeting their fundamental needs and thus cause harm. Denial of access to healthcare, housing, and employment through the use of AI perpetuates individual harms and generational scars. AI systems can kill us slowly.

Given what the "Gender Shades" findings revealed about algorithmic bias from some of the leading tech companies in the world, my concern was about the immediate problems and emerging vulnerabilities with AI that we could address in ways that would also help create a future where the burdens of AI did not fall disproportionately on the marginalized and vulnerable. AI systems with subpar intelligence that lead to false arrests or

---

* The Campaign to Stop Killer Robots was launched in 2013 to pass an international law to prevent the automation of killing humans. You can learn more on the official website: https://www.stopkillerrobots.org/.

wrong diagnoses need to be addressed now. The enthusiastic guest at the dinner, similarly, was concerned with addressing near-term problems with AI. They left dinner early, as they were on their way to advise on a biometric identity system in a foreign country. What if the enthusiasm to do good with AI ends up sending parachutes with holes on thankless and unwanted missions?

Looking across the table and thinking about so many people who would never be invited to a dinner like this, I doubted I would be receiving any future private dinner invitations. I thought about the excoded—people being harmed now and those who are at risk of harm by AI systems.

When I think of x-risk, I also think of the risk and reality of being excoded. You can be excoded when a hospital uses AI for triage and leaves you without care, or uses a clinical algorithm that precludes you from receiving a life-saving organ transplant.[6] You can be excoded when you are denied a loan based on algorithmic decision-making.[7] You can be excoded when your résumé is automatically screened out and you are denied the opportunity to compete for the remaining jobs that are not replaced by AI systems.[8] You can be excoded when a tenant screening algorithm denies you access to housing.[9] All of these examples are real. No one is immune from being excoded, and those already marginalized are at greater risk.

At this dinner, I realized my research could not be confined just to industry insiders, AI researchers, or even well-meaning influencers. Yes, academic conferences were important venues. For many academics, presenting published papers was the capstone of a specific research exploration. For me, presenting "Gender Shades" at New York University was a launching pad. Deserting the island of decadent desserts, I felt motivated to put my research into action, beyond talking shop with AI practitioners, beyond the academic presentations, beyond private dinners. Reaching academics and industry insiders was simply not enough. I needed to make sure everyday people at risk of experiencing AI harms were part of the fight for algorithmic justice.

PART IV

INTREPID POET

CHAPTER 13

AI, AIN'T I A WOMAN?

I sensed an opening. Research papers could reach academics and industry practitioners focused on AI, but I needed something more to reach everyday people. I also needed to reach decision-makers like elected officials who might be seduced by the promises of AI to bring increased efficiency without being aware of racial, gender, and other types of bias. Did the government officials in India exploring the adoption of the Aadhaar system know about the potential for bias in the biometric solutions being offered as answers for efficient distribution of government resources and persistent identification? Did they know algorithmic bias might deny benefits to the very people they sought to help? What about the police departments adopting facial recognition technologies? What did they know about algorithmic bias, if anything? I knew I couldn't leave it to the companies selling these systems to reveal their flaws. There was no incentive to put technological shortcomings in a sales pitch. I needed to humanize the harms and biases of AI systems and bring a perspective that tech companies were likely to shy away from. How might I use my knowledge to help people see beyond the headlines now being written about my work, "Facial Recognition Is Accurate, If You're a White Guy," and feel the impact on a specific person?[1]

I decided one way to humanize AI biases and make the topic more mainstream than an academic paper was to test the faces of the *Black Panther* cast. Since my research had shown that the systems I tested worked worst on the faces of darker-skinned females, I decided to focus on the faces of the women of Wakanda: Lupita Nyong'o as Nakia, Letitia Wright as Shuri, Angela Bassett as Queen Ramonda, and Danai Gurira as fearless General Okoye. I brought on Deborah Raji as my research intern to carry out a small-scale audit running the *Black Panther* cast's faces across the AI systems of five companies. This exploration became known as the Black Panther Face Scorecard project. The project revealed some commonalities with my own experience. Like me, some of their faces were misgendered, not detected at all, or in some cases mis-aged. Angela Bassett, who was in her late fifties at the time of the photo, was estimated by IBM's system to be between eighteen and twenty-four years old. (Maybe not all algorithmic bias was that bad.)

The results were amusing. The Black Panther Face Scorecard drew smiles from colleagues and visitors from member companies of the MIT Media Lab. These fictional characters, played by actors whose faces had reached billions of people, still felt safely removed from everyday life. While more women were rocking shaved heads, not many people were walking around with vibranium undershirts or bracelets with ammunition to keep superhero relatives safe. At least, this wasn't happening in my social circles.

The performance metrics on the women of Wakanda kindled my curiosity. How would these AI systems work on the faces of not just fictional dark-skinned women but iconic women of today and yesterday? How might AI read the faces of highly photographed women like Michelle Obama, Serena Williams, and Oprah Winfrey?

Screenshot of Oprah Winfrey image misclassification, from the visual poem "AI, Ain't I A Woman?" Youtu.be/QxuyfWoVV98?t=133.

And how would it do on historic figures like Sojourner Truth, who escaped slavery by buying her freedom and pushed for women's rights and the abolition of slavery? I was also eager to try the faces of Shirley Chisholm, the first Black congresswoman, and fearless journalist Ida B. Wells. I searched online for popular,

Screenshot of Sojourner Truth image misclassification, from the visual poem "AI, Ain't I A Woman?" Youtu.be/QxuyfWoVV98?t=39.

widely used images of these women, which Deborah Raji ran through systems that included IBM, Amazon, and Microsoft. When she shared the results, I was astonished.

Looking at just the names with the results in a spreadsheet was one thing. Seeing the faces of women I admired and respected next to labels containing wildly incorrect descriptions like "clean shaven adult man" was a different experience. I kept shaking my head as I read over the results, feeling embarrassed that my personal icons were being classified in this manner by AI systems. When I saw Serena Williams labeled "male," I recalled the questions about my own gender when I was a child ("Are you a boy or a girl?"). When I saw an image of a school-aged Michelle Obama labeled with the descriptor "toupee," I thought about the harsh chemicals put on my head to straighten my kinky curls, until I decided to embrace my natural hair. And seeing the image of a young Oprah labeled with no face detected took me back to my white mask experience.

For a while, I tried to remain detached from my research findings, which indicated that all systems tested worked worst for dark-skinned females. The research touched on other groups that also warranted attention, like darker-skinned males and lighter-skinned females. With the latest exploration of women I admired, I had an opportunity to bring dark-skinned women like me to the center stage. I had the power to put faces to what might otherwise be story-less silhouettes.

My first instinct was to create an explainer video like the one I made for the "Gender Shades" research paper. Doing that was familiar and comfortable. It allowed me to show some of the outrageous results from the position of an analyst explaining how the results reflected misogynoir, the term coined by Dr. Moya Bailey meaning the ways Black women, specifically, are insulted or discriminated against.

After writing the draft script for an explainer video on these iconic women, I showed it to a teaching assistant in a film class I visited periodically and asked how I could improve it. "What motivated you to work on it?" he asked me.

"The research paper is the beginning of a conversation, but the results are abstract. I do not want to subtract the humanity of the feeling of being misgendered, being labeled in ways beyond your control. I want people to see what it means when systems from tech giants box us into stereotypes we hoped to transcend with algorithms. I want people to bear witness to the labels and peer upon the coded gaze for themselves."

As I spoke, he nodded his head.

"Have you considered making a poem about this instead of a script?"

For years, there was a form of art I indulged in but kept largely hidden. I had notebooks and digital diaries filled with verses and phrases. Snippets of my poetry dwelled in shadowy places. I enjoyed writing, but it was mostly a private, vulnerable exercise: I'd intended to keep my poetry mostly to myself and a tight circle of sympathetic ears.

When the sunlight warmed me awake the next morning, the following phrase sat in my head, capturing how I felt about witnessing the cultural impact of Serena Williams, Michelle Obama, and Oprah Winfrey walking in their paths:

> *My heart smiles as I bask in their legacies*
> *knowing their lives have altered many destinies.*

As I brushed my teeth and looked into a fogged mirror, more words came into focus:

> *In her eyes, I see my mother's poise*
> *In her face, I glimpse my auntie's grace*

As I ruminated on the work more lines came to me:

> *Can machines ever see my queens as I view them?*
> *Can machines ever see our grandmothers as we knew them?*

My poem "AI, Ain't I A Woman?" was born. The piece held the emotions I had long suppressed. When I spoke the words of the poem aloud, my anguish and disappointment emerged. But for the full impact, the words needed to be paired with the images and disheartening labels that were slapped onto these iconic women by AI systems from leading tech companies. Part of what made the white mask demo more powerful than words alone was seeing me alter myself by donning a white mask to be made visible to a machine.

Until making the white mask fail demo, I thought of tech demonstrations as celebrations of what machines could do. If a demonstration included a failure, the demo gods had failed you. I thought of the way Steve Jobs, robed in a black turtleneck, not only talked about the possibilities of an iPhone but demonstrated the capabilities with carefully selected examples to tantalize onlookers and change the conception of what a cellphone could be. His words mattered, and so did seeing a simple gesture opening an application or switching screen views. Showcasing what his words meant completed the seduction. The Apple demos were a pathway into transforming existing beliefs about technology.

I was doing something similar but in the opposite direction. There were plenty of examples to show the possibilities of tech. I was collecting examples to show the limitations. My collection of failure demonstrations provided a counterpoint to the celebrations that accompanied technological advances.

The white mask failure I recorded was an example of what I call a counter-demo. But what exactly is a counter-demo countering? With the case of the white mask, I was providing a counter-narrative to the research and accompanying headlines lauding advances in computer vision. With "AI, Ain't I A Woman?" I decided to record screencasts to create counter-demos. These demonstrations countered the supposed sophistication of AI systems being eagerly sold. I assumed commercially sold products from these companies would perform fairly well on most people's faces if they were being sold to a wide market.

At the time, these companies had online demos of their AI product capabilities that were publicly available so anyone with some time, an internet connection, and a photo could upload an image and see how the demos worked. To make counter-demos, I screen recorded my visits to these websites and sat through loading animations of rotating wheels that preceded the display of results. Some included colored boxes that would be used to locate a head in an image. All had some type of description about what the uploaded images contained. When I uploaded an image of Sojourner Truth to Google's system, it returned the label "gentleman." Truth had fought to be treated on equal footing with a gentleman but was also vocal in saying that she too was a woman. Her famous 1851 "Ain't I A Woman?" speech inspired the name of my spoken word algorithmic audit. Truth was also in the business of sharing counter-demos to large audiences to demolish dangerous narratives.

IN THE 1840S, THE DAGUERREOTYPE BECAME THE FIRST PUBlicly available and widely used photographic process. Similar to how some may see artificial intelligence today, the daguerreotype was assumed to be objective, given the scientific tools used to develop photos. However, photography can be used in service of dangerous projects that hide under the mask of scientific objectivity. The slave daguerreotypes produced by Harvard's Louis Agassiz were developed to analyze differences between "African blacks" and "European whites" as a means of "scientifically" proving his view of white superiority. His daguerreotypes emphasized phenotypic differences in order to argue that there were multiple species of humankind, allowing for a racial hierarchy that put white Europeans at the top. The dehumanizing portrayal of largely undressed enslaved individuals in these "scientific" studies complemented the ongoing cultural and political denigration of Black people in the United States, justifying and naturalizing their subordination and brutalization. Aware of the power of images and the stories they can tell, Sojourner Truth used the power of photography to portray herself using the dress code associated with middle-class white women of the time. It was this image of Truth, intentionally wearing what were considered to be quintessentially feminine garments, that I submitted to Google's system—and that Google labeled "gentleman."

Sojourner Truth became a powerful orator pushing for abolition and women's rights while also pointing out contradictions in the arguments used by white women to justify these rights in her "Ain't I A Woman?" speech. She used her voice to push for change, but she also used her image to support herself financially by selling cartes de visite (collectible cards with photographs and messages that were a form of mass communication in the 1860s). Beyond providing financial support, Truth's images also provided support for ending slavery, joining the ongoing project of using photography to present Black people in dignified ways. Her images were her counter-demos.

She was not alone in this mission. Frederick Douglass used the power of photography to paint a different story with daguerreotypes than the one offered by Agassiz's slave daguerreotypes. Douglass became the most photographed person of the nineteenth century, and he used dignified portrayals of himself via daguerreotypes and subsequent photographic techniques to humanize Black people for the wider public as he pushed for abolition.

Truth and Douglass skillfully reused influential technology to shatter dehumanizing portraits that were constructed using the same tools. They showed that counter-demos do not just demonstrate but also demolish assumptions by offering real-world examples that shake the status quo. Similarly, AI systems can be used as tools of oppression as well as tools of liberation. Through counter-demo I challenge the concept of the tech demo as a performance that supports the adoption of technology. Like Sojourner Truth's critique of white women's marginalization of the perspectives and experiences of women of color—a critique that strengthens the impact of the women's rights movement—the critique of artificial intelligence is not a Luddite call to break machines, but a call to break harmful assumptions about machines and thus enable the construction of better tools and, more important, better societies.

WITH THE COUNTER-DEMOS RECORDED OF TRUTH AND OTHER iconic women, I then worked with the Ford Foundation to produce a video poem that paired the counter-demos with verse, creating the first algorithmic audit delivered as a spoken word poem. Building on the algorithmic audit that made up the "Gender Shades" research, "AI, Ain't I A Woman?" moved from performance metrics to performance arts. I more confidently called myself a poet of code, and followed the example of Truth and Douglass to make a counter-demo to oppose harmful racial assumptions. They used photography to counter harmful assumptions about the dignity of enslaved individuals. I used my counter-demos to combat assumptions about the neutrality of AI.

I am not alone in the use of counter-demo to show technical limitations. In 2009, YouTube user wzamen01 posted a viral video of an HP laptop with a face tracking feature. The video has received over 3 million views and more than sixty-five hundred comments at the time of writing. The video application shown was supposed to pan along with the movement of the face in the video stream. While the system worked fine for the person with lighter skin, in the frame referred to as "Wanda," for the darker-skinned person, the pan feature did not work. The person referred to as "Desi" states, "I'm Black. I think my blackness is interfering with the computer's ability to follow me." Seven years later, when I donned a white mask, despite the deep learning breakthrough and steady reports of technical improvement in the performance of facial recognition technologies, Black faces still broke the frame.

In September 2019, around six hundred thousand images were identified to be removed from ImageNet, one of the most influential computer vision datasets, in response to public scrutiny and criticism of offensive and derogatory labels used in the "person" category of the dataset. A now retired interactive tool called ImageNet Roulette, which was part of the Training Humans project developed by artist Trevor Paglen and scholar Kate Crawford, catalyzed massive public awareness. The ImageNet Roulette website enabled anyone with access to upload an image and have it labeled by a deep neural network trained on images from the ImageNet dataset. Some of the results were devastating, including counter-demos with instances of a dark-skinned man being labeled "wrongdoer, offender," an Asian woman labeled "jihadist," and more. The hashtag #imagenetroulette went viral. The website acted as a participatory evocative audit that allowed individuals to experience firsthand representational harms caused by an algorithmic system. This work demonstrates how a strategically positioned evocative audit can lead to real-world change. An army of individuals experiencing algorithmic harms shattered complacency, challenged the dominant discourse on machine neutrality, and furthered the push for combating algorithmic harms. Each social media post with the hashtags showing offensive labels created pressure on the makers of the ImageNet dataset to respond, particularly as news coverage about the viral hashtag spread. The contribution of the project was not only in changing ImageNet but also in elevating the conversation around the harms of developing and deploying algorithmic systems aimed at classifying people.*

FREDERICK DOUGLASS REMINDS US THAT THE STORIES WE evoke through imagery can allow those whose power does not rest on vast material resources to make change:

> Poets, prophets, and reformers are all picture-makers—and this ability is the secret of their power and of their achievements. They see what ought to be by the reflection of what is, and endeavor to remove the contradiction.†

---

* For a deeper analysis of counter-demos and the evocative audits they support you may want to read my doctoral dissertation: "Facing the Coded Gaze with Evocative Audits and Algorithmic Audits," dspace.mit.edu/handle/1721.1/143396.
† I viscerally encountered this quote at the "Vision & Justice" convening, as it was printed on the back of the curriculum compendium, which included my op-ed for *The New York Times*— "When the Robot Doesn't See Dark Skin" (July 21, 2018), www.nytimes.com/2018/06/21/opinion/facial-analysis-technology-bias.html.

As a poet of code, I paint pictures using words, performance, video, and technical research to highlight the contradictions in the promises we hear about technology—like artificial intelligence to advance humanity—and the reality I and others bear witness to when technology oppresses instead of liberates. With "AI, Ain't I A Woman?" I endeavored to create a piece that complemented the performance metrics of the "Gender Shades" algorithmic audit with performance art to viscerally convey the implication of the findings that showed for all systems they perform the worst for women of color. I let the poet take center stage, in the comfort of my computer screen. The real test was looming. How would other researchers and decision-makers receive this poetic risk? Would my research still be taken seriously? I was concerned that the subjectivity of my poetry would be viewed in opposition to the objectivity of my technical research. If it appeared that I already had a conclusion in mind before gathering and analyzing the data, I risked being considered a biased and thus less credible researcher. My future as an AI researcher was at stake.

CHAPTER 14

GATES IN BELGIUM

After a few wrong turns along seemingly identical cobbled streets in Brussels, Belgium, I found the meeting place. It was the summer of 2018, and I was eager to see if my art had a place in the halls of power. On this day I would convene with world leaders and executives of tech companies to offer advice on how to manage the pitfalls of artificial intelligence and deliver a gift. But first I had to get into the building.

I pushed through a set of heavy doors, each stamped with the emblem of the European Union (EU) and the EEAS (European External Access Service, the EU's diplomatic service). A guard stood behind a desk next to a set of metal detectors. I approached the desk and announced, "I am here for the EU Global Tech Panel convened by Federica Mogherini, the vice president of the European Commission."

He looked at me incredulously and said, "Federica doesn't just meet with anybody, and this is a meeting for important people."

Scanning the sweeping entrance hall, I felt out of place. Twenty-eight at the time, a woman standing no more than five feet two inches tall, I did not look like the diplomats in gray suits. To the guard I certainly wasn't the kind of person to attend a high-level meeting convened by a high-ranking EU official. I handed him the invitation letter I'd received. He barely glanced at it.

"Anyone could print anything off the internet."

He was not wrong.

Finally, I pulled out my phone, hoping the basic international plan would not fail me at this crucial hour, and I called Helene, Federica's right-hand woman. Her excited voice and warm demeanor were a welcome contrast to the icy reception. I explained I was having a bit of trouble and needed to be verified to be let in.

"I'll be right there."

As I waited, I started to look around the lobby to see what important people looked like. I also started to worry about time. I had arrived early to test out my newly created "AI, Ain't I A Woman?" video, ready for its debut. To my relief, Helene arrived to cosign for me. She firmly and politely told the guard to check the list. He finally asked, as he could have at the beginning:

"What is your name?"

"Joy Buolamwini, B-U-O-"

He quickly located it at the top of the list.

"Who do you work for?"

"I'm the founder of the Algorithmic Justice League."

He handed me my badge.

Just like algorithms confronted with individuals who do not

fit prior assumptions, the human gatekeeper stood in the way of opportunity, supposedly for the safety of those deemed credible enough or worthy to enter the building.

Unlike algorithmic systems, I could put a face to the decision-maker, and I also had the connections to challenge his initial decision. With black-box decision-makers, we are no longer facing the sexist hiring manager or xenophobic guard but a seemingly neutral device that nonetheless reflects the biases of the society it's embedded in. Unless we demand not only a choice in deciding whether and how these systems are used but also pathways to challenge decisions, we will change the form of the gatekeeper, but the prejudice will remain.

I cleaned my glasses in an effort to look more important and followed Helene to the gathering room. After plugging in my laptop to the A/V system, I tested out the room acoustics and the floor-based display screens that stood like pyramids in the middle of the room. Soon other members of the panel arrived, and I noticed I was the only one with a badge around my neck. I recognized one of the panel members who had also been at the dinner of deserted desserts. We sat in a circle and introduced ourselves. Throughout the meeting, I kept sliding my hand on the finger pad of my laptop to keep my gift to the group ready for action, hoping my battery would last. Toward the end of the meeting my fingers started slipping more as sweat gathered on my hands, and my chest began to rise and fall at a faster pace. I cleared my throat and addressed the group.

"Today we have gathered to discuss the threats posed by artificial intelligence and what the EU can do to be proactive. I am an algorithmic bias researcher and a poet: a poet of code. I believe it is important as we have these conversations to keep in mind how individuals are impacted by these systems. I have prepared for you all a spoken word algorithmic audit that shows how the faces of people like me who are largely absent from this room are read by AI systems. This is the first time I am showing this piece. My gift to you is its premiere. I hope it helps remind us all of what is at stake as we continue our work together."

My thumb hit the spacebar to start the video, and the face of a young Michelle Obama appeared on the pyramid screens. One by one, the members of my audience saw the captivating counter-demos. Some shifted their body weight when they saw the logos of companies they were connected with. When the video faded to black and the final note of the soundtrack rang, the last line of the poem hung in the air:

> *No label is worthy of our beauty.*

There was silence. I was glad I was wearing a blazer to cover the increasing sweat on my arms. Then there was applause and praise. Federica adjourned the official business, and we huddled in small groups. A man in a well-tailored suit walked toward me.

"That was a powerful video. It reminds me of a new research paper that came out called 'Gender Shades.' The paper recorded algorithmic bias from a number of different tech companies. . . ."

I interjected before he explained more of the paper to me.

"I am very familiar with that research."

"Oh."

"I am the lead author. It is based on my MIT master's thesis."

"Well, the paper showed these systems performed very well for white men but there were some notable gaps, especially with Black women."

I wasn't sure why he felt the need to explain the paper to me after I told him I had authored it. I was out of energy to keep challenging other people's perceptions of me.

"There is certainly work to be done," I replied.

Compared to other circles convened to explore issues around technology, the gender diversity of this group was one of the best I had experienced. But there was also work to be done on racial diversity. I worried that well-meaning panels like this would talk about issues with global implications from a largely Western viewpoint. To counter that tendency, the panel did include people from the Middle East and Northern Africa along with North American and European representatives. But besides the images in the "AI, Ain't I A Woman?" video, I was the only Black person represented in that room, though not the only first-generation immigrant. I felt an obligation to push for a greater representation of voices on the panel while also trying not to be impolite. I had barely made it past the guard. The experience reminded me I was a visitor to a strange land. I asked if there would be greater representation of the Global South in the future, venturing into the topic without pushing too hard. At the same time, part of being on a panel like this was to give voice to those not represented, as well as to talk about the implications of algorithmic bias research findings. The guard was not the only kind of gatekeeper to opportunity; all the members of the panel had gates of our own. What would we do with our influence?

Alongside gatekeepers are gate openers. I had been invited to be a part of the panel on the recommendation of Megan Smith, the former chief technology officer of the United States, the first engineer to hold the title as well as the first woman. Since a visit she made to the Media Lab, she had stayed in touch, though that was not my first encounter with her. She gave a talk at the Grace

Hopper Conference in 2009. At that time, I was an undergraduate studying computer science at the Georgia Institute of Technology, seated in the middle of a hotel ballroom in Tucson, Arizona. She was an executive at Google. With a cheerful voice, she implored us to debug tech's gender problems so more of us could be part of creating beneficial technology. Her enthusiasm and inviting attitude gave me a blueprint for talking about issues of diversity in the tech world. "We are problem solvers, and this is a problem we can solve if we take the time to focus on it," she encouraged. Listening to her speak made me want to be part of the change, and it also made me question what it was about her approach that I found motivating: I wasn't being chided.

During my trip to Brussels, I strolled into a comic book museum that appeared almost empty. On the top floor, I ran into a delegation from Rwanda. They were holding a reception. I looked as if I might belong, so I walked into the reception area with the confidence of a party crasher. There was no apparent gatekeeper and an unguarded table holding drinks. I approached a table and introduced myself. I spoke briefly about my work on algorithmic bias. I touched on the fact that police were using facial recognition technologies that had bias that would impact Black people already being overpoliced in the United States. I expected sympathetic ears.

"That is not really our concern. We don't have those types of African American problems in Rwanda."

I was reminded that some issues do not translate across a global context. Hearing my American accent, they probably didn't realize I was also African. I decided not to bring up the fact I was from Ghana in addition to being a United States citizen. Earlier I had been advocating for increased representation of views from the Global South, and this exchange reminded me of the prejudices African Americans faced outside of the United States.

In an attempt to connect on a different thread, I told them that my research used images of African parliament members, including those in Rwanda, to show issues of bias in technical systems. Still no interest. Instead, one of the men suggested, "Why not focus your energy on building something instead of critiquing what's not working?" They then shared an initiative they were leading that was focused on supporting the development of more scientists and innovators in Rwanda.

After my time in Ethiopia, I'd learned it was worth taking a pause to understand what was and wasn't working before building for the sake of building and the sake of innovating, though I thoroughly enjoyed making new things. Working on the Aspire Mirror project and exploring possibilities with computer vision and face tracking led me to focus on algorithmic bias. I was clear that I did not want to be in the business of building better facial recognition technologies that could then be used for increased surveillance.

Even if I didn't improve facial recognition technologies directly, in pointing out what was wrong my work did provide insights to those who were making these systems. Not long after my research was published, Quartz reported that Guangzhou-based startup CloudWalk was working with the government of Zimbabwe in an effort that would result in the collection of more faces from darker-skinned individuals. Data colonialism was here. Just like colonial powers had exploited the bodies and mineral wealth of Africa, now digital bodies were being extracted to build the wealth of foreign companies. Was I contributing to the ongoing exploitation of people like me?

In 2019 Google came under fire after it was reported that a

subcontractor coerced homeless individuals in Atlanta to surrender the valuable biometric data of their faces.[1] The identification of performance gaps using algorithmic audits can inspire predatory efforts to collect more data without regard for privacy or informed consent. The data collection efforts of CloudWalk and Google gave credence to the growing argument that my work was supporting surveillance. I even encountered one scholar who claimed that in "AI, Ain't I A Woman?" I co-opt figures who represent social justice aims, like Ida B. Wells, in a manner that strengthens carceral technologies. In my poem, asking "Can machines ever see my queens as I view them? Can machines ever see our grandmothers as we knew them?" was an invitation to further interrogate what kinds of algorithmic systems we create, not to condone carceral technologies. My goal was to stop harmful uses of AI and not bolster them, yet how other people used these research results or interpreted my poetry was not completely in my hands. What could I do to more actively resist harmful uses of AI, broadly speaking, and in the near term facial recognition technologies? What roles did individuals have in the fight for algorithmic justice when going up against powerful companies? How could I show that improving facial recognition systems was not the only response to understanding this research? My work was a warning shot and a necessary call to reconfigure and reconsider how we were designing, developing, and deploying AI products.

CHAPTER 15

POET VS. GOLIATH IN THE WILD

Snipers perched on rooftops, and the airspace over Davos was emptied as prime ministers, presidents, top-ranking government officials, and a poet of code entered the city. I had a feeling the snipers were not for my protection.

It was January 2019, and I was far from my research lab now. Just as I felt living at the corner of Bow and Arrow streets as a resident tutor at Harvard, in this space I felt the sense of being at the intersection of privilege and oppression. Like my experiences at the Media Lab, where I stood out against the pristine white walls of the buildings, I stood out against the white-encased landscape of the Swiss Alps in winter. Months earlier, I had received an invitation to present my research at the World Economic Forum (WEF), an annual gathering of political, business, cultural, and other leaders aiming to set global and industry agendas.

I was uncertain about attending, and not just because of the bitter cold. WEF was a symbol of globalization and free trade, which increasingly concentrated world power in the Global North and corporate hands. I reached out to Ethan for advice, as he had been selected as a WEF Young Global Leader a decade prior. The promise of access to powerful decision-makers was undeniable. Yet there were other power dynamics to consider. Would my participation really make a difference, or was my inclusion solely for virtue signaling? In the end, Ethan advised that you don't make change by talking only to the people who agree with you. He cautioned me to adjust my expectations, since there were many different tiers of access. Finally, he gave me tips for navigating the Swiss transportation system and suggested I get a bigger coat.

Bundled in a red oversized parka, I went to the registration area and received a white badge with my face printed on it. I walked down a covered tunnel past guards with guns. Equipped with the right access credentials, I was directed to a container housing an unusual piece of conference swag: attachable snow spikes for shoes. Apparently broken ankles were not uncommon for the ill-equipped, so I picked up three pairs. I was not going to test my luck on the icy streets. I walked through the complex to find where I would be presenting my research and showing my video poem. As I made my way around treacherous ice patches, often disoriented and feeling out of place, it was clear to me that I was not part of the inner sanctum of WEF. When a guard reached for a gun as a friend tried to drop me off at a hotel for designated badge holders, I was reminded that my inclusion in certain spaces was the exception to overall exclusion. On my way there, a friendly face called my name and waved enthusiastically. It was Professor Cynthia Breazeal, who was part of the delegation accompanying MIT president Rafael Reif to the forum. She told me about her experiences her first time attending and gave me some pointers on making the most of my time. I felt grateful, and reassured, to be walking in her footsteps yet again.

When I started my art project as a master's student, I never imagined the fallout would lead me to a stage in Switzerland talking about the shortcomings of AI, as world leaders explored ways to use AI in what was being termed the fourth industrial revolution. I was not exactly sure where I fit in. Was I the gadfly issuing warnings? Was I the junior scholar providing technical insights to give companies a competitive edge? Was I the poet to provide provocative words and entertainment? After shedding my initial pre-talk anxiety and delivering my presentation, I was eager to get home, but I left the audience with a teaser. I announced, "In just a few days, I am publishing another research paper as a follow-up to 'Gender Shades.' It includes results from all the companies shown here and another tech giant. Watch this space!"

That evening, I went to a reception hosted by MIT. There I saw President Reif, fundraising. I was not used to seeing those at the top of the food chain in my environment needing to solicit support. Intellectually I knew one of the roles of a university president was to raise money, but I had not seen or heard it in person. "These are the rooms where it happens," I thought to myself. I understood that in this space, my role was to be a representative of why contributing to MIT was a good investment. Was I a show pony? I was invited to another MIT event the following day, but I had other plans. I was free. I decided to be the party girl. The extra snow spikes did not go to waste as an elegant pair of piano-black dress shoes and another pair of tennis shoes announced the arrival of two travel buddies who were here to celebrate my twenty-ninth birthday. Playfully they had taken photos of themselves with Happy Birthday signs in swim gear while standing in the snow on the back porch framing the zigzagging mountain landscape. As I blew out the candles on a rainbow-bedazzled cake, I made a wish. "Give me love, grace, and protection. May I be a vessel for change."

Ethan and Megan had advised me to take advantage of my time in Switzerland to build connections. I had other ideas. I decided to take a full day off to celebrate my birthday and make memories with my friends in the snow. We started with snowboarding lessons.

In addition to the empty airspace over Davos, the ski slopes were largely abandoned. According to our snowboarding instructor, many locals left Davos during WEF and opted for the extra rental income and peace away from the influx of visitors. He patiently inched us down the practice hill before encouraging us to take the ski lift and try coming down the mountain with our shaky basic moves. As a former skateboarder, I was humbled each time I found myself seated on a throne of snow. Eventually, I got into a halting flow, picking up speed, sliding and slipping down the pristine white-powdered mountain. My overconfidence with my snowboarding abilities given my prior experience skateboarding was not unlike how some tech companies were launching AI products. Success in one area does not guarantee success in another. Using skateboarding techniques to attempt to snowboard left both my pride and my bottom sore. Launching immature and inappropriate products also affects the long-term bottom line of tech companies. Using skewed datasets or relying on assumptions that necessitate ideal conditions also risks failure. Skateboarding and snowboarding offer lessons for the development of artificial intelligence. Just like having an AI system learn on training data, the experience gained from snowboarding on a practice hill is valuable, but it does not fully prepare you for navigating steeper terrain. And just as with skateboarding, you might start learning at a park designed with obstacles and ramps thoughtfully spaced apart, but that experience does not fully prepare you for street skating, which in my case meant navigating the sidewalks of suburbia, where features like pebbles, sidewalk cracks, and potholes were not designed for skating. Bruises followed.

AI approaches that showed promise with creating efficiencies, such as optimizing the use of energy in data centers, are tempting to port into other domains. After it was acquired by Google, DeepMind proved useful in saving the company 40 percent on the cooling cost of its data center after using AI to optimize energy efficiency.[1] Optimizing a fairly constrained environment is not quite the same as working with an unconstrained environment. Early explorers of face detection techniques soon ran into the limit that the training data of people photographed in well-lit environments with set poses did not equip systems for unconstrained environments, also known as the real world, where most people do not walk around with studio-quality lighting. When researchers began collecting images of faces "in the wild," candid images of people posted online to improve face detection systems, this was like moving from a nice indoor skate park with a beginner's area and into the streets to navigate real-world conditions.

Mountain snowboarding employs different ratings to show the difficulty and risk for certain routes. Beginners have no business trying to navigate Double Black Diamond routes reserved for those with not just confidence but experience. Applying AI systems that have not been tested on a wide range of people or conditions to high-stakes scenarios is like a beginner confidently going down a Black Diamond route before learning the basics and without the proper equipment. As I was snowboarding in Davos that January, we still lived in a world where law enforcement could adopt AI systems that had not been externally tested for accuracy to inform investigative leads. False arrests followed, as we would see in the cases of Robert Williams, Michael Oliver, and Nijeer Parks. All these Black men were arrested due to facial recognition misidentification.[2]

While I navigated WEF, in the back of my mind I was preparing myself for another event. Deborah Raji, in addition to working on the Black Panther Face Scorecard, wanted to work on a research publication during her internship with me. We decided to replicate the "Gender Shades" paper and see how companies were doing a year later—and we decided to add two new companies. She recommended we add Clarifai, where she had done an internship. I recommended we add Amazon, since they were providing facial recognition technologies to police departments and were the subject of multiple resistance campaigns. In the prior summer, a hundred and fifty thousand individuals and forty civil rights organizations had signed letters demanding the company stop selling their Amazon Rekognition product to law enforcement.[3] Law enforcement applications of any emerging technologies were already Black Diamond territory, given the stakes. Using facial recognition technologies that hadn't even been shown fit for the intended purpose yet, let alone tested externally, was both dangerous and irresponsible.

Deborah, whom I started to call Agent Deb, and I were scheduled to present our findings at the AAAI Conference on Artificial Intelligence, Ethics, and Society in January 2019 in Honolulu—right after WEF. The day before the conference, *The New York Times* was scheduled to release a feature on the front page of the business section about algorithmic bias findings. The process of finalizing the article was not smooth. At the last minute, Amazon representatives claimed they had not had a chance to see the results, though I had documented evidence of sharing this information with the company. Maybe the information got lost or was sent to spam. In addition, the previous summer I had posted a public letter to Jeff Bezos, with similar preliminary findings.[4] Now with a peer-reviewed paper being published alongside a *New York Times* feature article, Amazon could no longer ignore the research findings. Based on Amazon's behind-the-scenes pushback and the time it would take me to get to the AAAI conference from Switzerland, on my last day in Davos I wrote a Medium article anticipating the pushback from Amazon. Two thousand-plus words of counterarguments flowed from my fingertips as I typed with intensity on my laptop.

I sent an email with the rebuttal article to dozens of reporters who had covered the "Gender Shades" paper earlier and let them know that *The New York Times* was about to drop an article on the new research paper. I let them know that I would not be available to comment for twenty-four hours because I would be traveling overseas. If they needed quotes there were plenty in the Medium post. I hoped that would be enough for the time being. I needed rest.

I arrived in Honolulu in snow boots and a big red winter coat, surrounded by people in Hawaiian shirts, sandals, and leis. The AAAI conference organizers had the good sense to choose a tropical location for a January conference. As much as I had enjoyed the snow slopes, Waikiki Beach was a welcome reprieve from the winter bitterness of Davos and Boston. Agent Deb and I met in a hotel lobby. She was set to present the next day. Since she was first author on the paper, I insisted she present it alone. It was not the typical practice to have an undergraduate as the lead author of a paper, but I wanted to show that young researchers could make meaningful research contributions. I would be there to support her just like Timnit was there to support me on the "Gender Shades" paper the prior winter.

Agent Deb walked through her first-draft presentation with me. I raised my right eyebrow when she finished. "You can't present it like that. . . . They will laugh at us." She winced a little. We sat on the floor for the next few hours and revised the slides. I was especially hard on Deb because I knew that as young, marginalized researchers going up against some of the largest tech powers in the world, we would not be spared. I would rather be hard on her now so we could withstand what was to come. Of the things I could teach her, I emphasized that what she should learn from me is how to communicate and share research in a way that gets attention.

"Many people can gain tech skills. Your ability to communicate and tell a story about your technical work is what will separate you from your peers. It's what makes people see why change is necessary. I don't want to hurt your feelings. I want you to be ready for the headwinds."

The next day, I left my items in my seat and walked around taking photos like a proud parent as Agent Deb presented "Actionable Auditing" to a room of about four hundred people. Francesca Rossi from IBM was there, and I had been sitting with her and some of her colleagues. In this paper, IBM had done significantly better, and Amazon trailed behind all its peers. The "Actionable Auditing" paper received an award for best student paper, and I stopped taking photos to go pose with Deb for the official conference photographer. Deb's first research publication, on which she was the first author, had received an award and a cash prize. We also had our photos in the *New York Times* article that came out: "Amazon Is Pushing Facial Technology That a Study Says Could Be Biased," by Natasha Singer.[5] My promise to Agent Deb when she joined me came true: "I cannot pay like a tech start-up, but what I can teach you will set you up to go beyond me. Learn from all of us and be better."

THEN CAME THE ATTACKS. AN AMAZON VICE PRESIDENT, DR. Matt Wood, claimed our paper and the *New York Times* article were "misleading" and drew "false conclusions."[6] In particular, the issue Hal had raised years before resurfaced. In short, because our paper focused on the gender classification feature of Amazon Rekognition, Wood argued that those results did not apply to the facial recognition capabilities of the Rekognition product. I thought I had been clear in stating that the research results raised concerns about other face-based tasks. Did they genuinely not see the connection? Unlike the reception I received from IBM, which made me feel change was possible, Amazon's approach made me question the extent to which large corporations could be trusted.

At the time Amazon and Microsoft were both gunning for a $10 billion contract to provide AI services to the Pentagon.[7] This was not a great time to have your AI capabilities questioned. Amazon's counterattacks continued, and I wrote a second Medium article to address their claims. I felt like I was in the wilderness single-handedly fighting the Goliath known as Amazon. None of the other companies had taken a combative approach. I wondered if I had taken the research a step too far by challenging Amazon. I worried I might have put Agent Deb—an undergraduate with aspirations of going to graduate school—in too much danger. Prospective computer science departments might per-

ceive her as too much of a risk. Future employers in the tech industry could blacklist her. She asked me what she could do to
help as the attacks mounted. "Let me handle this. Focus on your
schoolwork." I kept her in the dark as I strategized what to do
next.

As the attack from Amazon was happening, I was heartened
that organizations like the American Civil Liberties Union of
Massachusetts and the Georgetown Law Center on Privacy and
Technology came to our defense. They posted spirited messages
on Twitter to rally behind both the research and me personally.
We needed more. I was concerned that if other researchers saw
what Amazon was doing to us and no academics stood up in defense, other researchers would perceive the professional risk of
this kind of research as being too high. The research community's response to Amazon's attack on "Actionable Auditing" would
set a precedent. I told Timnit and Meg Mitchell my concerns. At
the time they were the co-leads of the Google AI Ethics team and
in a position to speak out. They organized a letter signed by
seventy-five researchers, stating their support for our research.
Signatories included Professor Anima Anandkumar, the former
principal AI scientist at Amazon, and Professor Yoshua Bengio, a
winner of the Turing Prize (I think of it as a Nobel Prize for computer science). All these academics working in the computer science space took significant professional risk by standing up for
the research, given how much research funding comes from Amazon. Bloomberg ran an article titled, "Amazon Schooled on AI
Facial Technology by Turing Prize Winner." Yoshua captured the
sentiments of the researchers' letter when he said that Amazon's
response to the study was "disappointing." He also said, "It is important to have the social and scholarly debates about what is so-
cially and ethically acceptable in the use of these new technologies.

This case highlights such issues in a very clear way and is a good way to increase public awareness." I felt grateful for the support from my allies beyond my host institution.

Shortly after the "Actionable Auditing" paper came out, Amazon announced they would work with the National Science Foundation (NSF) on funding the Program on Fairness in Artificial Intelligence, the very area Amazon had publicly attacked. Yochai Benkler observed in an op-ed for *Nature:*

> When the NSF lends Amazon the legitimacy of its process for a $7.6-million programme (0.03% of Amazon's 2018 research and development spending), it undermines the role of public research as a counterweight to industry-funded research. A university abdicates its central role when it accepts funding from a firm to study the moral, political and legal implications of practices that are core to the business model of that firm. So too do governments that delegate policy frameworks to industry-dominated panels. Yes, institutions have erected some safeguards. NSF will award research grants through its normal peer-review process, without Amazon's input, but Amazon retains the contractual, technical and organizational means to promote the projects that suit its goals.[8]

I agreed. I also noticed that the public support I received was largely outside of MIT and the Media Lab. I was especially grateful to Kade Crockford of the ACLU of Massachusetts and Alvaro Bedoya of the Georgetown Law Center on Privacy and Technology, who publicly defended my work on social media. I reached out to Ethan: "I need backup! I am going up against Amazon."

Ethan got into gear and wrote a post defending my work.

Others I reached out to had various reasons why they could not support me publicly. They were mainly concerned about funding or antagonizing a company as influential as Amazon. I felt like I had been largely abandoned by the majority of MIT leadership at that moment. When "Gender Shades" was receiving positive public attention, there was no hesitation to stand with me in the sunshine. Now in the midst of negative attention, would they stand by me? I was seeing for myself the influence of Amazon. Amazon supported major research initiatives, and MIT and the Media Lab were in fundraising mode. This was probably not an ideal time for them to ruffle feathers.

In December of 2019, I excitedly called my parents. "I've been vindicated by the feds!" The National Institute of Standards and Technology had finally released a long-awaited paper about the effect of race, age, and sex on the performance of facial recognition software. According to the government website, "For one-to-one matching [facial verification], the team saw higher rates of false positives for Asian and African American faces relative to images of Caucasians. The differentials often ranged from a factor of 10 to 100 times, depending on the individual algorithm." I was surprised to see that the difference in accuracy could be up to 100 times worse for Asian and African American faces as compared to Caucasian ones. The findings also revealed: "For one-to-many matching [facial identification], the team saw higher rates of false positives for African American females. Differentials in false positives in one-to-many matching are particularly important because the consequences could include false accusations."[9] So while Amazon was right that my studies had looked at gender classification and not facial verification or facial identification, I was correct in my assertion that the bias observed in my studies was cause for concern in other areas, given shared technical approaches used with a whole range of facial recognition technologies. Unlike Microsoft, Amazon did not submit its systems to be tested for this landmark study in 2019. By the time NIST's vindication came in December, my illusion of safety had already dissolved. I had assumed basing my work at an academic institution like MIT would offer me backup and support if my work was challenged. I assumed the Media Lab, which showcased my work to attract students and news coverage, would defend me. It was disappointing to feel like I had to plead for protection and then receive very little. I was even more eager to leave the lab at every opportunity I got. But if the university was not the place I could speak truth to power, where could I go? Who would have my back, and who could I unite with?

CHAPTER 16

BROOKLYN TENANTS

Several months after I left Hawaii with Agent Deb and did all I could to defend my research, I felt unanchored. Something was missing. Talking about the dangers of facial recognition technologies and AI harms was only a starting point. There was more to this work than just theory or research. There was more to the work than talking to tech companies. Yes, the companies needed to change how they developed and deployed AI products. And those changes could help prevent future harm. But there was no guarantee companies would change without legislation. I wanted to be closer to the ground. I wanted to help real people who were being impacted by these systems. I wanted to feel like my work mattered, not because it was validated by other academics but because it made a valid impact in the lives of the excoded.

. . .

MY OPPORTUNITY TO CONNECT DIRECTLY WITH THE EXCODED
came in April 2019. Scrolling through my inbox, I slowed down
when I saw a time-bound request from a Brooklyn lawyer. I had
less than a week to respond.

Friday, April 26, 2019

Hello Joy,
I work with the Tenant Rights Coalition at Brooklyn Legal
Services in NYC, and we're working on a case involving a
private residential building owner attempting to install facial recognition software in lieu of the current keyless fob
system at a large residential building in Brooklyn. Your publication, *Gender Shades,* raises serious accuracy and bias concerns about the proposed system, given that the tenants of
this large 700+ unit complex are predominantly people of
color, women, and elderly.

We were hoping you could help us better understand the
potential for algorithmic discrimination with the type of
system the building owner has proposed to install here. The
facial recognition software is manufactured by StoneLock,
which uses the heatmap of an individual's face to allow
them access. They claim it does not differentiate between
gender, sex, or color because it "merely reads data points on
a person's face and assigns a number." I suppose our very
basic question is, *can the facial recognition system proposed here
still discriminate based on demographic and phenotypic characteristics even though it uses heatmap technology and not video-*

surveillance facial recognition? We find it puzzling that they can guarantee there is no risk of algorithmic discrimination when zero validation studies exist on the accuracy/bias of this particular system. I've attached copies of a marketing handout and a paper explaining the proposed entry system that the building owner shared with us. Any thoughts or insight would be much appreciated.

Also, this is a bit of an urgent matter. The tenants live in a rent-stabilized building and so, before it can replace the current entry system, the landlord must seek approval from the state agency that oversees rent-stabilization laws in NY. We're working with the tenants to file an opposition with the state agency by next Wednesday, 5/1, and are hoping you can share some technical guidance as soon as possible so we can strengthen our argument against the owner's application.

Look forward to hearing from you!

They were working on an opposition statement against a landlord's application to install a facial recognition entry system at the Atlantic Towers apartment complex. After taking a preliminary call, I reviewed their documents and gave some quick feedback on the parts that I understood, beyond the legalese. As I was boarding a flight to North Carolina on the way to visit friends, I received a call from the lawyer to talk about my feedback. More calls followed as she scrambled toward her deadline, and she had a second request.

"I hope this is a good time to call."

"I'm in the airport, but we should have a little time."

"Would you be willing to work on an amicus letter of support for us?"

"Uh, I will need some time to think about it."

I needed some time to look up what an amicus letter of support was, and I had to decide if I had any business drafting such a thing. I was straying further outside the lines of typical graduate student work, spending less time at the Media Lab and more time on trains and planes to share my research findings. Increasingly, it was only Foodcam images of leftovers in my email inbox that reminded me that I was still in school. I had finished my doctoral coursework, and my supervisor was fairly hands off. I took full advantage. Like my first year as a master's student, I felt free to explore. Instead of focusing on publishing more papers from my thesis work, I put my attention on reaching people outside of academia. "AI, Ain't I A Woman?" was being incorporated into a number of art exhibitions around the world. I had even shipped the white mask that started this journey as part of a five-year traveling art exhibition curated by the Barbican Centre in London, opening May 16. The amicus letter was needed by May 12. I had also been summoned to offer expert testimony at a congressional hearing on May 22. I had just enough time and more than enough motivation.

I tackled the amicus letter first. After reviewing StoneLock's marketing materials and the opposition statement submitted by the tenants with the support of the coalition, I started working on a statement. Each public statement from AJL served two purposes: one, to voice support of opposition toward particular actions, and two, to educate others who were considering either adopting or opposing a similar system. I felt excited to be able to directly use my research and knowledge to support the tenants. Hal's rhetorical questions about whether this work mattered no longer bothered me. I could see firsthand the hope it gave to people facing real-world harms. The nagging drive to continuously prove myself also began to fade. The research mattered not because it was presented at an academic conference or featured in *The New York Times*. It mattered because it could be used to make palpable change in the lives of everyday people.

To support the cause, I used StoneLock's own materials to highlight the need for people using a security system to be in support of the system.

> Security operators recognize systems that are transparent to the user as having a distinct advantage over systems that introduce annoyance or frustration. The fact is: *user adoption and compliance, both conscious and unconscious, is vital to the success of any security system.*[1]—StoneLock

One hundred and thirty-four tenants at Atlantic Towers had voiced strong frustration with the StoneLock system in the opposition letter. They were absolutely justified in their concerns about the accuracy of the system, consent, data breaches and exploitation, and security risks. They worried that algorithmic bias would mean the system would fail on their faces, making it difficult to access their home without additional hassles. They had not explicitly agreed to have their faces used for a biometric entry system when they started living in the apartments. They had no guaranteed protections that prevented their face data from being sold and used for other purposes or handed over to the police. In the amicus letter I provided research and relevant examples to back up what the tenants were already saying. StoneLock's marketing materials provided no information about how their system performed on different types of faces. This was especially concerning because more than 90 percent of the tenants were people of color. They were predominantly woman-identifying, and they included minors as well as the elderly. The vast majority of the tenants belonged to one or more of the groups that have the highest failures in U.S. government–sponsored studies I checked that examined the accuracy of facial recognition technologies.[2]

StoneLock claimed that they had tested their system in 40 percent of Fortune 100 companies. I doubted the company demographics matched the demographic of the tenants at Atlantic Towers. The tendency to test systems in one context and then transport them to another is an invitation for context collapse. As tempting as it may be to rely on initial tests in one environment to justify deployment in another, the real world is more complicated. I remember hearing Kate Crawford share the line that "caribou are not kangaroos." At an AI Now event held at the Media Lab, she said that self-driving car systems trained in a country where caribou crossed the road would be ill-equipped for Australia, where instead of a caribou crossing the road, kangaroos took their chances. However, when a kangaroo hops, a system expecting an animal to move straight across the street might erroneously pump the gas and run into the kangaroo on its way down. Fortune 100 employees are not Atlantic Towers tenants.

Another example of context collapse occurred when Winterlight Labs, a Canadian start-up, created a system that used machine learning to attempt to detect indications of Alzheimer's disease from voice recordings.[3] Their system had been trained on Canadians who spoke English as their first language. When the system was tested on Canadians who spoke French as their first language, these speakers did not match the training data. The context collapse occurred because the signs being used to infer

Alzheimer's disease may also overlap with signals conveyed by someone searching for the right word or stringing together words in unusual ways in a second language.

When I met with Tranae Moran and Icemae, longtime tenants in Atlantic Towers, they had many questions about data. We sat at a picnic table between towering buildings. Icemae asked, "How do we know they aren't giving our data to the police?" and "Can't our data be hacked?" Tranae wanted to know, "If they are keeping my face data shouldn't I get some money?" and "Is our face data useless like they claim?" Their questions demonstrated that people being impacted by algorithmic decision-making were far from uninterested once they became aware of the risks. These women were reminding me of the personal impact of algorithmic harms. The work was far deeper than a research project. The responsibility I felt to help became heavier. What started as a graduate art project was now being used to show deployments of facial recognition technologies gone awry.

Tranae and Icemae were right. Ongoing efforts of tech companies to collect more diverse datasets reveal that their data as Black women was in demand. As I was working on the "AI: More Than Human" Barbican art exhibition, I was seeing another space where a lack of diverse data appeared. Deepfakes were on the rise. A technique known as generative adversarial networks (GANs) allowed for the creation of photorealistic faces of fake people. Yet these fake people were seeded by a dataset of training images. And as for all machine learning systems, data is destiny. The curators for the exhibition wanted to display images of people from the Pilot Parliaments dataset in the exhibition. However, the EU's new General Data Protection Regulation had a provision saying that redistribution of biometric data required consent. I didn't want to take any chances since the dataset had the faces of EU citizens in the form of parliament members from Iceland, Finland, and Sweden. I expressed my hesitation to the Barbican.

The exhibition construction team said that they could just show the faces of the Africans in the dataset. Even though there were no laws protecting African people, I did not want them to have fewer protections, so I objected. As a solution they proposed using a deepfake face generator to represent the dataset. I warned that "these systems are likely trained on skewed data, so I suspect the image of those meant to represent Africans might not look quite as realistic as others." They went for it anyhow. When I reviewed the datasets, the lighter-skinned deepfakes seemed like plausible humans, while the darker-skinned ones had many examples that simply could not pass for a photorealistic representation. The hairlines and texture seemed to represent no human—and not in a good way.

The elderly and predominantly Black and Brown tenants were not the faces that dominated the most prolific publicly available face datasets at the time. Furthermore, the demographic and phenotypic makeup of the training set used in StoneLock's proprietary system was unclear. The company was asking us to simply trust their word without showing how the system was trained. When it comes to AI systems, trust must be earned and not assumed. Tranae and Icemae had every right to be concerned about police being given access to their data. In 2019, there was no federal law that explicitly addressed the use of facial recognition technologies, leaving residents at risk of grave harm. There was no law that explicitly provided oversight of law enforcement or government use of facial recognition technology.[4] Law enforcement and even federal agencies like the FBI and ICE could seek access to the system's valuable store of biometric data linked to personally identifiable information. This access could expose an already vulnerable community to targeted harassment and worse. Such an arrangement would make tenants vulnerable to police profiling and false accusations, in addition to further data exploitation and privacy breaches.

Furthermore, although StoneLock argues that near-infrared light can improve accuracy over other techniques, they do not address the specific challenges of near-infrared facial recognition. For example, the accuracy of near-infrared facial recognition may be affected by the emotional and physical condition of the individual, which can be influenced by illness, alcohol, or exercise.[5] StoneLock's corporate training grounds, workplace environments where alcohol consumption is limited if not banned, and which employees tend not to frequent when ill, may prove ill-matched to the variability introduced in a residential community.

StoneLock also attempted to claim that their system was bias free because of the type of imaging they used. Yet they did not have adequate data to support this claim. Instead, they suggested that the infrared technology they were using shielded them from the concerns associated with other facial recognition systems.

EVEN IF THE STONELOCK SYSTEM WAS SUFFICIENTLY ACCURATE for the tenants, the visible light face data requested by outside authorities could be used with other facial recognition systems, introducing new risks. No person should be required to submit their face data to law enforcement or immigration officials in exchange for a roof over their head, but in an unregulated climate, there is a significant risk the information could leak from the owner's hands into the government's.

Once I had the amicus letter drafted, I circulated it to a number of authors and academics who were increasingly vocal about the harms of AI. Cathy O'Neil signed on. Meredith Broussard, author of *Artificial Unintelligence,* said yes. So did Dr. Ruha Benjamin and Dr. Safiya Noble, the authors of *Race After Technology* and *Algorithms of Oppression,* respectively. Dr. Sasha Costanza-Chock, who was at the time working on their book *Design Justice,* also signed. In my conversations with Sasha, they emphasized how important it was that communities facing the brunt of issues design their own response. Sasha's conceptualization of design justice invited me to reconsider my role as a researcher as not being in front of those who were being harmed but rather being alongside and at times behind. Lived experience was just as valuable if not more so in certain contexts than academic expertise. The Brooklyn tenants led the charge. They were already aware that something was not right and were actively leading protests and resistance campaigns. I gathered champions of the Algorithmic Justice League to help strengthen their cause. Time would tell if the landlord at the Atlantic Towers would back away from the plan to install the facial recognition entry system, at least temporarily. Without laws governing facial recognition technologies, any victory would be tentative because the decision could be reversed. Still, it was a contribution worth celebrating, a contribution that showed that resistance to face surveillance is possible with direct organizing. I had found another way to fight for algorithmic justice outside of ivory towers and private dinners. Supporting people on the front line with research and preparing accessible informative materials were absolutely necessary in the movement for algorithmic justice. Though it wasn't required for my academic studies, all the extra work of putting together a companion website and explainer video made a difference where it truly mattered to me: outside the lab. Tranae told me:

"After finding [this] work, it just lit me up on fire, I was like . . . these are the facts right here. There is a clearly thought out website. It was very easy to share the ['Gender Shades' findings] with my neighbors because it was very digestible, it was language that we could all understand. It just helped me share what I had learned with my neighbors who are mostly elderly people."*

WHILE I COULD NOT DIRECTLY CONTROL THE ACTIONS SECUrity and tech companies took to improve facial recognition technologies, I could and did focus my efforts on equipping others. Educating people on the front lines of the fight for algorithmic justice was something I saw as another vital role for AJL.

In addition to being used by impacted communities directly to resist algorithmic harms at a local level, the "Gender Shades" research has been incorporated into successful efforts to push for legislation that restricts the use of facial recognition technology by government agencies, including police departments. After the research was published, Matt Cagle and Jacob Snow of the ACLU of Northern California drew inspiration from the academic work to do an algorithmic audit of Amazon Rekognition. The ACLU audit following the "Gender Shades" design focused on elected officials. Instead of international parliaments, the ACLU audit focused on the U.S. House of Representatives. Their aim with the

* For a discussion of the *Coded Bias* movie, see www.facebook.com/codedbiasmovie/videos/819183838883059/. Tranae's quote can be heard at 21:34.

"Gender Shades"–inspired audit was to raise public awareness about police use of facial recognition and move lawmakers to enact legislation. Twenty-eight members of Congress in their audit were matched, incorrectly, with mugshots.

Moreover, campaign materials developed in support of restricting police use of facial recognition referenced the "Gender Shades" research to provide evidence for valid concerns, including the fact that these tools exacerbate already existing racial bias. On May 14, 2019, San Francisco became the first city to ban police use of facial recognition. The ACLU of Massachusetts has also led successful legislative campaigns in coalition with a number of organizations that have made Massachusetts, as of this writing, the state with the highest number of municipal restrictions on government use of facial recognition. These Massachusetts campaign efforts also incorporated the results of the "Gender Shades" findings.[6] I was delighted and a little caught off guard by the reach and impact of the work.

These experiences showed me the power of collective action and the benefits of taking time to go outside the lab. The critiques of my work that pointed out how it could be used for companies to shore up shortcomings of their AI systems are true. The "Gender Shades" focus on classification accuracy can strengthen state power by reifying categories of control and providing insights into areas where control can be subverted, hence enabling the state to strategize on how to mitigate or thwart subversion. At the same time, large errors in classification, using contestable labels, demonstrate how fallible AI systems—even those produced by leading companies—can be. Such errors disrupt facile assumptions of machine neutrality and technological objectivity.

It is also true that the same work was used to support bans of and moratoriums on harmful uses of facial recognition technologies. Being quiet about my findings would not have prevented harm, because these systems were already in development. My speaking out provided an opportunity to consider alternative pathways, including nonuse. This was the case with San Francisco's prevention of law enforcement use of facial recognition. Following the release of "AI, Ain't I A Woman?" Google stopped using gendered labels in their commercial AI systems, showing that awareness about bias could lead to changes in the design practices of a company. It was not a foregone conclusion that gender labels like "gentleman" had to be used in their automated labeling at all. Other companies started using the less definitive descriptions of their work and opted for "perceived gender" instead of "gender" in describing the kind of labels their systems produced. I am more convinced than ever of Audre Lorde's words: "Your silence will not protect you." Despite the mixed responses to my work, I am glad I spoke up. But the movement needed—and still needs—more people inside and outside of labs speaking up when they see harmful AI systems. We need AI practitioners, when seeing a lack of representation in datasets, to use their position to document the issues and make sure the limitations of a system or research findings are published alongside the hopeful possibilities. We need employees willing to block product releases or change the design of a system to prevent harm.

We need artists who use their creativity to craft evocative pieces that humanize the impact of AI-driven mistakes. We need tenants who speak up against the installation of intrusive systems and take the initiative to learn more about the technologies we are often encouraged to accept without questioning. We need researchers who take the time to make work accessible and digestible so the greater public knows what is at stake and advocacy groups can take up that work to support successful campaigns.

We also need to acknowledge the risks associated with speaking up against powerful entities. In the early days of my work, I was largely shielded from direct threats. Those days did not last when my research challenged a tech giant. We needed a counterweight to corporate power.

CHAPTER 17

TESTIFY

"Thank you, Chairman Cummings." I glanced over at Max, Megan Smith's black-and-white cat. He looked uninterested but was nonetheless involved in my practice run before my first congressional testimony. Max was quite focused on licking his right paw as he unfurled near the swimming pool.

BY MAY 2019 I WAS FEELING MORE DISTANT FROM ACADEMIA and more motivated to connect with people harmed by AI systems. Yet, I had more academic hoops to jump through. I needed to do my general exams, a series of tests from my committee to assess my mastery of my area of research. The exams included written take-home assignments. But I was barely home these days, as attention to my research brought opportunities to do art exhibitions and speak at conferences all over the world. When I received an invitation from the House Committee on Oversight and Reform to testify at a congressional hearing, I was ready to try my hand at persuading policymakers to prevent AI harms. Testifying at the hearing was not just a matter of showing up to answer questions; I also needed to submit written testimony ahead of time. How was I going to write three general exams and prepare for this hearing? My overflowing inbox, untouched meals, and perpetual exhaustion were markers on my road to impending burnout. I reached out to Ethan to see if my real-world written testimony could be a stand-in for one of my general exams. For all my frustrating experiences while in grad school, I could not deny the level of flexibility and freedom I had as Ethan's student. He agreed. Ethan had also introduced me to the Ford Foundation in 2018, and in May 2019, the foundation became the first organization to fund the Algorithmic Justice League. The money from the Ford Foundation and the reduction of my workload renewed my energy. AJL was officially now more than an idea, and I was more than a student.

BACK BY THE POOL, I WAS REFINING THE WORDS OF MY OPENing statement and getting feedback from Megan as we lounged in her backyard in the home nicknamed the Embassy of Innovation. "Keep it bipartisan," she advised. "Make sure everyone on the committee understands how it impacts their constituents." On my flight from Boston to Washington, DC, I had put the final touches on my written testimony, which had to be submitted in advance of the first of a series of hearings on facial recognition technology. Twenty-four hours before the hearing, there was still time to craft my opening five-minute statement.

Earlier that day I had been part of a moot hearing arranged by Alvaro Bedoya and Laura Moy at Georgetown. Alvaro explained that *moot hearing* was the jargon they used to describe a practice hearing. We were in a conference room that had been reconfigured to have two rows of tables facing each other. One table was for the interrogators and the opposing table was for the witnesses. I looked over at my fellow witness, who seemed calm and prepared. For me, the practice round had not gone well. The first question Alvaro directed my way proved to be a stumbling block. I muddled my way through an unsatisfactory response and wondered what business I had giving congressional testimony. My face grew hot. I seldom blushed, but today I was purple. The stakes were high: This hearing could push the federal government to pass much-needed federal legislation on facial recognition technologies. Alvaro reassured me that the questions he asked during the moot hearing were likely to be much harder than anything I would face at the real hearing. "It will only be broadcast on C-SPAN," I told myself. "Who watches C-SPAN?" Then Ethan texted me: The Media Lab was arranging a viewing party in the third-floor atrium.

Alvaro's kind words of reassurance did not stop me from studying my notes for hours, anticipating questions that fit the theme "Facial Recognition Technology and Its Impact on Civil Rights and Liberties." I printed out about twenty pages of notes and reviewed them every free moment I had. The pressure was building. I would be testifying alongside Neema Singh Guliani, senior legislative counsel with the American Civil Liberties Union; Clare Garvie, a coauthor of "The Perpetual Line-Up," a report that helped push me to study algorithmic bias; Andrew G. Ferguson, a professor of law; and Dr. Cedric Alexander, the former president of the National Organization of Black Law Enforcement Executives. Neema was a regular at congressional hearings and responded with sharp, crisp answers and an air of "I said what I said" confidence. I was asking around about the proper way to address the committee members and other basic questions so as not to embarrass myself. At least now I knew the meaning of a moot hearing. Maybe I had finally flown too close to the sun, but it was too late to back out now. If I was going to go down in flames, at least it would be in style. My look was ready: red blazer, white-rimmed glasses, braided faux hawk, and a bracelet I had selected for its resemblance to the Wakanda kimoyo beads shown in the film *Black Panther*. One thing I was certain about was that I would need supernatural powers to make it through the next day with my dignity intact.

At 7:00 A.M. on May 22, 2019, I woke up in the playroom of the Embassy of Innovation. I kneeled beside old toys and science projects. "God help me! Give me the words to say. Give me strength, courage, and wisdom." Today was not a day to lean on my own understanding. The final version of my opening testimony was freshly printed downstairs. I bounced down a set of stairs with Max underfoot to join Megan at the kitchen table. I practiced my remarks in between bites of bagels and cream cheese.

At 11:00 A.M. Chairman Elijah Cummings, a Maryland Democrat from the district that includes the city of Baltimore, began the congressional hearing. In his opening remarks he said, "I can relate to facial recognition mistakes. I cannot tell you how many people stop me, thinking I am John Lewis. Sometimes, I don't have the heart to tell them, and I just let them take a photo." Ranking member Jim Jordan, a Republican from Ohio and a vocal Trump supporter, also gave remarks focusing on the privacy implications of the use of facial recognition. Over the duration of the hearing, he would return to the point that these systems were being used without the approval of Congress. "How does the FBI get access to this data?" He scowled. In a highly polarized time in American politics, with the 2020 elections looming the next fall, we had managed to find a topic that concerned both parties. The Democrats were especially focused on the civil rights implications and the fact that Black and Brown communities were at even greater risks of harm from facial recognition. Chairman Cummings spoke about the deployment of facial recognition on protestors in Baltimore who attended a rally to condemn the killing of Freddie Gray at the hands of law enforcement. In addition to touching on police brutality, he noted the chilling effects of deploying facial recognition at protests—a threat to the First Amendment right of freedom of expression. Would you feel free to attend a protest if you knew law enforcement would deploy facial recognition to record who attended?

The Republicans often returned to the privacy implications, in which a Big Brother–like government could watch people's movement. Glenn Grothman, a Republican representative from Wisconsin, brought up a point I had not considered. "As we begin to have politically incorrect gathering places, a gun show or something, is it something we should fear that our government will use it to identify people who have ideas that are not politically correct?"*

Clare Garvie responded with a clear voice of authority. "Law enforcement agencies themselves have expressed this concern. Back in 2011 when the technology was really getting moving, a face recognition working group including the FBI said face recog-

* House Hearing on Facial Recognition Technology, May 22, 2019, C-SPAN, www.c-span.org/video/?460959-1/house-hearing-facial-recognition-technology. Glenn Grothman's comments appear at 01:55:28.

nition could be used as a form of social control, causing people to alter their behaviors in public, leading to self-censorship and inhibition."

A common assumption about lawmakers is that, when it comes to technology, their knowledge can be limited at best. Throughout the hearing, I was pleasantly surprised to see that the lawmakers and their aides had clearly done their homework. Democratic representative Jimmy Gomez from California confessed,

> I must admit, I was not even paying attention to this technology until I was misidentified last year during the ACLU test of members of Congress. And it really did spark an interest and a curiosity of technology and really did feel wrong deep in my gut. I started looking into it. . . . I've had nine meetings with representatives from Amazon, we've asked questions from experts across the spectrum, and my concerns only grow. . . . Despite the fact that Amazon had not submitted its product to outside testing, it still sold that product to police departments. . . . Ms. Buolamwini, do you think third-party testing is important for safe deployment of facial recognition technology?

Was a congressman really asking about third-party testing of AI systems? "Absolutely," I replied. "One of the things we've been doing at the Algorithmic Justice League is actually testing these companies where we can. . . . We absolutely need third-party testing and we also need to make sure with the National Institute for Standards and Technology that their tests are comprehensive enough."

"Yes, because if it's evaluating on a dataset that is incorrect or

biased it's going to lead to incorrect results,"* Representative
Gomez responded before yielding back to the chairman.

After a while, I realized that not all lawmakers were there to
ask questions. Given different schedules, representatives came in
and out of the hearing, some staying long enough to make remarks that would provide good sound bites for campaigns, others
coming in from overlapping committees. Partway through the
hearing, representatives Alexandria Ocasio-Cortez from New
York, Rashida Tlaib from Michigan, and Ayanna Pressley from
Massachusetts, three members of the group known as the Squad,
joined. The Squad represented a new wave of junior members
who espoused progressive left-leaning views. When Representative Pressley spoke, she paused to note that the Algorithmic Justice League had been started in a city in her district, Cambridge,
Massachusetts. She related that she had read about the concept of
the coded gaze in my recent *New York Times* op-ed, and then she
gave me the floor to explain the term to the committee.

Representative Ocasio-Cortez then directed her questions
toward me. Poet of Code (POC) and AOC locked into a flow.

"Ms. Buolamwini, I heard your opening statement. . . . We
saw that these algorithms are effective to different degrees. Are
they most effective on women?"

"No."

"Are they most effective on people of color?"

"Absolutely not."

"Are they most effective on people of different gender expres-
sions?"

"No, in fact it excludes them."

* House Hearing on Facial Recognition Technology, May 22, 2019, C-SPAN, www.c-span.org/
video/?460959-1/house-hearing-facial-recognition-technology. Jimmy Gomez's comments appear at 02:10:00.

"Which demographic is it mostly effective on?"

"White men."

"And who are the primary engineers and designers of these algorithms?"

"Definitely, white men."

"So we have a technology that was created and designed by one demographic, that is only mostly effective on that one demographic, and they are trying to sell it and impose it on the entirety of the country?

"So we have the pale male dataset being used as something that's universal, when that isn't actually the case when it comes to representing the full sepia of humanity.

"And do you think it could exacerbate the already egregious inequalities in our criminal justice system?"

"It already is."

Chairman Cummings interjected in a matter-of-fact voice, "How so?"

Turning toward him, without skipping a beat I responded, "So right now, because you have the propensity for these systems to misidentify Black individuals or Brown communities more often, and you also have confirmation bias, where if I have been said to be a criminal then I am more targeted."*

Our back-and-forth captured a simple truth. The privileged few were designing for the many with little regard for the harmful impact of their creations. The consequences continue to ripple. Michael Oliver, Nijeer Parks, Robert Williams, and Randall Reid were all arrested due to misidentification aided by automated facial recognition. Williams was wrongfully arrested in

* House Hearing on Facial Recognition Technology, May 22, 2019, C-SPAN, www.c-span.org/video/?460959-1/house-hearing-facial-recognition-technology. Comments appear at 1:48:52–1:50:17.

front of his two young daughters and held by law enforcement overnight.[1]

Even though Williams was eventually released, the memory of his two young girls seeing their father arrested as neighbors looked on cannot be erased. He shared that he put his children in therapy.* The trauma of being falsely arrested—and knowing resistance could end in a fatal situation—is indelible. These are just some of the cases we know about. I think about the inmate who sent me a desperate letter from behind bars and others we may never hear from or who may never know facial recognition technologies had a hand in their arrests.

The ray of hope remained this: We were still in the early days of the creation of facial recognition technologies and artificial intelligence. Our actions today will have generational consequences.

Chairman Cummings concluded the hearing, stating, "In my twenty-three years, this is one of the best hearings I have seen. Really, you all were very thorough. Very, very detailed." Glancing at the clock, which indicated nearly three hours had passed since the beginning of the hearing, he said, "Again, I want to thank all of you for your patience. . . . This meeting is adjourned."[†] With that, he slammed down the gavel. Then, he beckoned me to approach the summit of the rows of seats behind wooden panels that led to his perch. When I reached the top of the congressional Everest, he leaned in, looked me in the eyes, and said, "I promise you, Congress will do something about this."

---

* In 2023, the Algorithmic Justice League presented Robert Williams with the inaugural Gender Shades Justice Award to recognize his advocacy efforts. He testified at a 2020 congressional hearing on facial recognition and continues to speak about the dangers of facial recognition. www.ajl.org/gender-shades-justice-award.
† House Hearing on Facial Recognition Technology, May 22, 2019, C-SPAN, www.c-span.org/video/?460959-1/house-hearing-facial-recognition-technology. The full remarks appear at 2:57:11.

The grueling hours of testimony had galvanized the leader of the oversight committee. He leaned in closer and whispered in my ear, "Now, what do you advise that we do first?" I was honored that he was seeking my input and also a bit surprised. I thought my job was over. I realized my role was not only to educate but also to suggest concrete actions that were achievable and impactful. I paused to think about the least controversial action that could be taken that would help everyone involved in the debate about the future of facial recognition technologies take steps to reduce harm.

"Chairman, we need a baseline. Right now we do not have a complete picture of how the federal government is using facial recognition or how organizations receiving federal dollars are using it. At a minimum, the committee should commission a survey so we know where the technology is being used, to what purposes, and which companies are or have been selling to all government agencies."

"We can get that started."

I then pushed for something that would take more time but also underscored the action that I thought would reduce the most risks and harms for now.

"Given all the risks and threats, I strongly advocate we put a moratorium on government use of facial recognition. Let's take the precautionary principle so we are not deploying technologies where there hasn't been adequate scrutiny and debate and where mistakes are very costly." He nodded his head and shook my hand, saying, "Thank you, Ms. Buolamwini."

I thanked him for the opportunity to offer my perspectives. His aide gave me a big smile and got up from her chair. "I am also from Ghana. You make us so proud. We will stay in touch, and we want to learn more."

I had not flown too close to the sun. Leaving the Rayburn Building, I looked to the sky and smiled to myself. I was born to fly with words and testify with conviction.

Two more invitations to testify at congressional hearings that summer followed. In June, I was back at the Embassy of Innovation practicing opening statements with Max the cat. I declined the third invitation, because I had to prepare for my August oral examination to become a PhD candidate. I double-dipped and convinced Ethan to let both of my written congressional testimonies count for two of my written exams.

The night before my oral exam, my body reached a point of exhaustion I had seldom experienced. I felt faint and even walking was challenging. I felt my heart speeding up unexpectedly as if I were being chased. I tried taking deep breaths, but this made no discernible difference in calming my nerves. The back-to-back travel, the exhibitions, and the testimonies had proven too much to handle in such a short amount of time. In desperation, I went to an emergency room near my apartment, wondering if I should cancel the exams. Draped in a hospital gown, I thought about the questions my committee members would ask me. I did not want to dwell on my health. After hours of waiting and a few examinations, I was discharged in the morning with a recommendation to slow down. "We see too many stressed graduate students," a nurse told me as I packed my belongings. A few hours later I logged into a video conference call to face my examination committee. Somehow I managed to pass. I was making steps toward this terminal degree. I felt I was walking in my purpose, but I knew I needed to take better care of myself. I needed all my strength.

Congress was interested in action. I received follow-up questions from both sides of the aisle. Issues about the harmful use of facial recognition technologies did not belong to just one political party. The more lawmakers who could identify dangers and risks associated with emerging artificial intelligence, the more likely we could get necessary legislation passed.* The success of our hearing was followed by a steady stream of lobbyists visiting different lawmakers, according to reports I got from my colleagues in Washington. Laura Moy and I, as we considered subsequent actions, kept returning to resource constraints. With our small organizations we simply did not have the same amount of time, staff capacity, or money to continuously visit representatives the way the lobbyists can. It was time to go back to the streets. How could we pressure Congress to move and buffer against lobbyists protecting the interests of businesses and not everyday people? How could we show millions of people what was at stake with bias in AI?

---

* In the UNESCO book *Missing Links in AI Governance*, I co-authored the chapter "Change from the Outside: Towards Credible Third-Party Audits" with Deborah Raji and Sasha Costanza-Chock. The chapter provides policymakers with recommendations for preventing algorithmic harms. www.unesco.org/en/articles/missing-links-ai-governance.

CHAPTER 18

BETTING ON CODED BIAS

"**W**here are my keys?"

"Not here," said my glasses case.

"Try later," said my coat pocket. I still had not mended the hole in the lining.

The timing could not have been worse. Another important development in my fight for algorithmic justice began to take shape about a year earlier, in 2018, when Shalini Kantayya's name kept appearing in my direct messages on social media platforms and in multiple inboxes. She had seen my TED-featured talk and was interested in interviewing me for a film project. I looked at her background and discovered she had directed *Catching the Sun*, a climate justice film executive produced by Leonardo DiCaprio. Scrolling through my Netflix account, I noticed the film was available on the platform. I then watched her TED talk and learned we were both part of the Fulbright program. I decided to write her back. She quickly arranged a flight to Boston to meet me. I needed to get a good night's rest. The next day I was set to do an on-camera interview.

My search for my lost set of house and office keys proved futile. I called a friend, seeking refuge. I caught a ride to meet her, and she walked me to the back of her manicured house. She ushered me into a well-organized guest bedroom and took a moment to examine my hips. In the morning she dropped off a fresh pair of underwear.

"Don't worry, I haven't worn these yet." She laughed.

"I'm guessing I don't need to bring these back."

While I could change some of my undergarments, I was out of luck with my other clothing. The outfit I planned for the shoot was locked away in my apartment. I wore the same clothes and glasses I had the day before. I spread oil on my nails to substitute for a finishing coat and showed up to my office keyless. Fortunately, given my habit of losing my keys, I was acquainted with how to have the facilities group or Media Lab communications director Alexandra Khan assist me. Alexandra was the key to making sure Shalini and future crews, photographers, and journalists had all the access they needed. Today, she had the master key to open my office.

Just in time for the crew to set up, my office door was opened. Shalini and I sat down for one of many interviews and conversations to come. Steve, a tall, stocky man with black hair pulled into a bun, operated the camera and asked me to pause between responses so he could make adjustments. My office was infiltrated with lighting equipment and reflectors. After the interview, Shalini shadowed me as I gave a presentation for a lab meeting at the big Civic Media table just outside my office.

After seeing me in my professional context, she made a request I would hear often: "Can I capture you doing something normal? Can I film you making coffee?"

"I don't drink coffee."

"Making tea?"

"I prefer not to be shown eating or drinking on film."

"When people see a film they aren't going to connect to the research; they are going to connect to you as a person."

"You can film me selecting my glasses."

I was hesitant to show more of myself beyond my role as a researcher. As an extremely private person—although you might not be able to tell from this book—the last thing I wanted was a filmmaker following me around in my personal space or in intimate moments, especially not a filmmaker I had just met. I didn't think my personal life was that interesting. Half my time was spent looking for lost items. But I did consider what she was saying. Shalini wanted people to see me as someone relatable. To make the process work I would have to trust her and open up a little bit more. Still, I worried. What if she cast me in a negative light? With some hesitation, I thought about what she might show of me outside the lab. I settled on having her film the process of having my hair styled at a salon full of Black women: Simply Erinn's. I wanted her to see who I spoke to outside the lab about my research and also where I drew a sense of broader community beyond academic circles.

As she asked the stylists awkward questions about the Black hair business, I wondered if I had made the wrong choice. I smiled sheepishly at the stylists, hoping to convey that she was safe to talk with. Shalini was especially brave. Over our months of shooting, I witnessed her show up in many different types of spaces with me to tell an evolving story with no clear end in sight. The process would not always be comfortable, but being willing to go through those awkward moments and missteps was necessary. My step to being more open on-screen started in the stylist's chair, my hair as unfinished as the story Shalini was spinning. In that chair, I had no laptops or screens to hide behind. If the aim was for people to see me as more than a researcher, we would start at the roots. If I did believe the heart of computing was humanity, I would need to get a bit more acquainted with showing my own humanity and get over the fear that I would lose credibility and respect if I was not always polished. With half my hair out in an Afro and the other half being skillfully twisted by Ranae, my go-to stylist, I continued the conversation on algorithmic bias. Under the roar of a hair dryer, I shouted over to Shalini, "I hope this is intimate enough for you."

By the summer of 2019, like so many independent documentaries, the film was running out of funds. Shalini accompanied me to Atlantic Towers Plaza, but before filming the Brooklyn tenants and me, she revealed the lack of money to record more footage. I attended a convening where I spoke to a communications professional brought specifically to help advise nonprofits. I told her about the ongoing production of the documentary, my excitement about the topic but also the need for additional support. Given all the work before me and the opportunities on my plate, she looked at me with pity and asked: "Are you sure this documentary is really worth your time?"

Was she right? Was I wasting my time? It had been around a year and there was no plausible end in sight. I shot back, "It will be worth the time if we make the investment to support it."

She was not the ally I had hoped she would be in that moment. The moment gave me clarity that I was indeed taking a risk by allocating my energy to the uncompleted documentary. Before there was anything to show for it except expense reports, I believed the risk was worth it because Shalini was capturing important stories about algorithmic harm that needed to be told. Her documentary, unlike more well-funded endeavors about issues with the tech ecosystem, centered on people of color. What was apparent to me was not so apparent to others. I had to believe in myself in the face of doubt. The purpose of the film was to expose the coded gaze, and I needed to take a bet on Shalini, on myself, and on the gravity of what we were depicting. I hoped that this film, if it ever got completed, would powerfully humanize the impact of AI bias and make the harms of AI a conversation for many people to partake in.

Unfortunately, if we wait to see immediate impact before we invest time, talent, treasure, and networks into supporting creative endeavors, they may never have the support they need to thrive. At a point where the film's coffers had reached their lowest, I connected with Doron Weber of the Alfred P. Sloan Foundation. Doron Weber had the foresight to make a $50,000 writing grant to Margot Lee Shetterly, who wrote the book *Hidden Figures,* which later became the uplifting film. I had learned of this backstory when on the Rhodes Scholar mailing list I offered my reflections on the film and my work on algorithmic bias. I remembered attending a film screening at MIT where Margot gave a talk. Seated in a bouncy blue chair at the Kendall Square Cinema, I was moved to tears when I saw the Black women walk together toward greater opportunity. I was surprised that seeing technical, analytical, and brilliant Black women on-screen had so much emotional impact on me. I had become accustomed to seldom having others who looked like me doing the kind of work I explore.

I left the *Hidden Figures* screening room with a deep understanding that representation is vital and telling stories that are often hidden or seldom elevated not only corrected the historical record but also enabled others to see within themselves the possibility to make contributions broader society would say were beyond their reach, scope, intelligence, or capability. Intellectually, I understood role models made a difference. *Hidden Figures* took that understanding from an intellectual acknowledgment to a heartfelt experience. We are capable, we are essential, then and now. Months later I received a copy of the book. Looking at the first few pages, I found a handwritten message: "Congratulations on winning the search for *Hidden Figures* contest, and thanks for accepting the torch from these women." Margot's signature completed the gift, the echo of her hand passing to me the flame of encouragement.

Doron and I had connected on a mailing list a few years earlier, so I reached out and let him know the film was running short on funding but running high on impact. He agreed to meet with Shalini. I made the introduction while we were visiting the Brooklyn tenants in the spring of 2019, shortly after leaving the Atlantic Towers complex. It was now up to Shalini's persistence and persuasiveness to convert the warm lead into resources.

Though there were countless rejections along the way, the Sloan Foundation came through, providing essential funds to help get the film to completion. Funds are necessary but not enough. The film itself still had to be finished, and the frenetic dash to the end could have been a film all its own. Right before Thanksgiving of 2019 Shalini told me the good news: The film had been accepted into Sundance! At the time I didn't know the difference between Sundance and Moonpie; I just knew Shalini was extremely excited. Over the phone, she explained, "Joy, this is the film festival where the leading studios and distributors come to choose films to purchase. A Sundance premiere gives a small documentary like ours the best chance of reaching a large audience."

"It sounds like I should make plans to go. Where is it held?"

"Park City, Utah."

"When is it?"

"January."

Before we could prepare our snow boots, there were a few remaining hurdles. The film did not have a name. Neither of us was thrilled with the working title *Code for Bias*, but more pressingly the film did not have an ending. I told Shalini that the Brooklyn tenants had made significant progress in resisting the installation of a facial recognition entry system by their landlord. I sent her links to the related headlines, including "Brooklyn Landlord Does an About Face on Facial Recognition Plan" and "How We Fought Our Landlord's Secretive Plan for Facial Recognition—and Won."[1] As Shalini decided to continue that storyline, how the film would end would remain to be seen. I was neither the director nor a producer on the film. The film would be distributed by Women Make Movies and be sold by 7th Empire, Shalini's production company. My behind-the-scenes efforts to support the film financially and convince my collaborators to participate did not give me control over how the story would be told. The Algorithmic Justice League would receive no proceeds from the film nor were I or any of the film's cast compensated to be in the documentary. This was a major labor of love.

Documentaries are generally made over many years, not in the month-to-month sprints we had just undertaken. To make the documentary ready for Sundance required double overtime from all involved. Shalini and Steve with the bun came to Boston to shoot final scenes. We arranged a video hangout call with Tranae and Icemae. I wrote a poem I hoped to share with them to honor them for their courage and example.

Unlike the first time Steve and Shalini came to Boston, I had my keys and plenty of clothes. I put together several outfits, gathered a few pairs of glasses, got the AJL shield, and polished my recently acquired double monk-strap dress shoes. Then, I made my way to the shoot location, a coworking space I was renting for AJL. When we arrived, we raided the open kitchen, with snacks set aside for us. We would not have to spend as much on lunch today. Switching rooms in the coworking space, we had enough variety to give the appearance of having traveled to multiple locations. We reset cameras and lighting equipment on different floors, mindful that Shalini and Steve had a flight to catch that same day. I can say there was much to be grateful for in November 2019, but pushing borrowed production equipment from the shoot back to the MIT Media Lab was not at the top of my list. The three-block walk was exhausting. On my way, I waved to Shalini and Steve, who zipped by me in a car in a last-minute attempt to make their early-evening flights. I had not signed up to be part of the production crew, but by this point everyone was pitching in however they could to reach the finish line. The wheelchair ramp on the side of the Media Lab was a welcome sight; it eased the burden of getting the equipment back to my office. Time would tell if the pain in my back and the year-end frenzy were worth it.

The Sundance Festival began on January 23, 2020, my thirtieth birthday. Several years after first coding in a white mask, my journey from the lab to the halls of Congress, from graduate student to founder of the Algorithmic Justice League, would soon face the spotlight of the documentary world. Striding onto the red carpet with the crew and cast, we took photos wearing black rectangle sunglasses that resembled censorship strips to symbolize resisting face surveillance. I was starting to feel ever-increasing eyes on me, but I was not entirely comfortable in the spotlight. I decided not to dress up too much as I felt uneasy being the main protagonist of the film. Thinking about the negative reactions I had already received with the TED Talk, I wondered if a film like this would only amplify the haters. On the other hand, I thought about so many people who had reached out, telling me that my work made them feel seen. The spotlight both shines and burns, and tonight my job was to focus on the shine. Our delegation eagerly walked into a completely full theater to welcome the premiere of what we ultimately titled *Coded Bias*. I watched both the film and the audience. I felt elated as the crowd cheered on a slow-motion sequence showing me buckling my shiny shoes, slipping a red ring on my pointer finger, and topping off the look with the AJL shield strapped to my back. I saw people leaning forward in their seats to take in stories of people I had come to call the excoded.

The excoded are individuals or communities harmed by algorithmic systems. The excoded included people like Daniel Santos, a schoolteacher who received low rankings from an automated system despite winning numerous "Teacher of the Year" awards. The audience was completely silent as he said, "For a moment I doubted myself . . . then I realized the algorithm is a lie." By the time the film finished, audience members were full of questions for Shalini and me about what they had just witnessed. The energy felt similar to what I had experienced at the *Hidden Figures* screening years earlier. There was a sense of awe and pride. I felt incredibly lucky and privileged that we had been able to help make a film that moved people to question assumptions about the capabilities of AI. At the after-party in a basement bar, I found Doron seated at a booth in the back corner. Reaching out to congratulate me, he said, "This is one of the best investments we've ever made in a science documentary." I wondered if he said that about all his documentaries. I decided to take the shine and give him the benefit of the doubt.

A year after the Sundance worldwide premiere, *Coded Bias* was broadcast nationally on PBS through *Independent Lens*. The television station that had carried my nine-year-old imagination into MIT now showed me at MIT discussing artificial intelligence. Maybe there was another child seeing me and imagining new possibilities. Shortly after the PBS debut, *Coded Bias* was released on Netflix to the more than 200 million subscribers of the platform at the time and translated into thirty languages. The film received critical acclaim and numerous awards. I truly appreciated the recognition, and I especially cherished messages from a wide range of people:

I am a 66 yr old white man, and I have been concerned about racism for decades. In the last 4 years I have become appalled at the realization that we live in a White Supremacist society that seemingly has no bounds. Your program "Coded Bias" further enlightened me into the depths that prejudice continues to be pursued against my brothers & sisters of color. Learning that face recognition is simply a new frontier that continues and promotes this evil practice left me saddened and very concerned. I wanted to say "thank you" for uncovering the truth of this while Algorithmic Bias is still in relatively early stages. Your work is extremely important and I am grateful and amazed by your brilliance in discovering and pursuing this injustice.

I am literally in tears as I watch the Netflix documentary. Your intelligence, bravery, and cultural pride makes me so happy. I obviously want to thank you for research and your auditory voice, but I also want to thank you for the spoken word, WuTang earrings, and the prideful cultural exhibition of braiding your hair. Thank you!

I just finished watching *Coded Bias,* with tears in my face. Happy that the world has such freedom fighters like you. Your drive and determination for the work you and your cause [do] is truly an inspiration. As a physician, I am part of the diversity, equity and inclusion committee for one medical society and one medical device company. As the owner of a research company, collecting data on patients for medical use and prediction, your documentary opened up my mind to potential mistakes I could have made as we will soon be starting predictive analytics. You are brave!

I just watched *Coded Bias* with my 14 year old stepdaughter who loves math. She was blown away by the power of algorithms and the importance to insert ethics in it. I have tears of joy seeing the amazing work you and the AJL is doing.

Letting my guard down on camera if even just partially is a risk I now see as having been worth it. I keep the messages I receive to motivate me when I find myself in moments I want to give up. The response to the film reminds me of the impact of Pete Souza's photo of the young boy touching the head of President Obama. *Coded Bias* provided an opportunity to create powerful depictions to reach people considered outsiders in the tech world. Spotlighting my journey and that of so many others, the film shows that marginalized voices matter. That the tech industry needs us. Policymakers need us. And we need each other. As a researcher wanting to reach beyond academic circles, I had to embrace working and collaborating with skilled storytellers like Shalini to shatter broadly held assumptions we have about technology, artificial intelligence, and whose voices and stories count, whose perspectives are deemed worthy of depicting. As women of color we lifted each other up and took a chance when there was no clear payoff and already mounting costs.

On September 29, 2022, I stepped out onto a humming New York City street and looked up at the marquee of the Palladium Times Square. "Welcome to the 43rd Annual News and Documentary Emmys" beamed down on me. After picking up my ticket, I went down the entrance escalator escorted by a dear friend in a handsome tuxedo. His pocket square matched my red draping dress. To my delight, the first person I saw on the red carpet was Shalini, glowing in gold. When our eyes met she waved me over to join her. We embraced, barely containing ourselves as we said to each other, "We did it!" *Coded Bias* was nominated for the Outstanding Science and Technology Documentary category. While I would have liked to win, the impact of the film and the millions of people reached was a reward no golden statue could touch. *Coded Bias* showed me the importance of investing my time in creating media that offers perspectives that might otherwise be marginalized and storytellers who are often overlooked. I thought about meeting Tranae and Icemae at a time when I was looking for meaning outside of academic work. I thought about the people I would never know but who would be moved by their stories. I no longer doubted if I could use story-telling in the movement for algorithmic justice. After years of hesitation, I found my voice as a poet of code.

As I left Times Square, I reflected on words I wrote for Tranae and Icemae, well before an Emmy nomination or a Sundance premiere was a possibility.

TO THE BROOKLYN TENANTS

To the Brooklyn tenants
Resisting and revealing the lie
That we must accept
The surrender of our faces
The harvesting of our data
The plunder of our traces
We celebrate your courage
No Silence
No Consent

You show the path to algorithmic justice requires a league
A sisterhood, a neighborhood,
Hallway gatherings
Sharpies and posters
Coalitions Petitions Testimonies, Letters
Research and potlucks
Dancing and music
Everyone playing a role to orchestrate change

To the Brooklyn tenants and freedom fighters around the world
Persisting and prevailing against
algorithms of oppression

automating inequality
through weapons of math destruction
we stand with you in gratitude

You demonstrate the people have a voice and a choice.
When defiant melodies harmonize to elevate
human life, dignity, and rights.

The victory is ours.

PART V

JUST
HUMAN

CHAPTER 19

DROP OUT

Simone Biles headed into the 2020 Olympic Games as the global face of gymnastics and the crowd favorite to win all her events.* I took pride in seeing her dominate her sport and was looking forward to seeing her collect another set of gold medals that seemed destined to grace her neck. The stage was set for her to shine and for me to cheer her on in my living room. Eschewing international expectations, she pulled out of the All-Around Final event, citing the need for safety and a case of the *twisties*. Far from the Ariake Gymnastics Centre in Tokyo, I followed along as sports journalists and former gymnasts explained the twisties, a dangerous phenomenon that occurs when an acrobatic athlete can no longer register how her body is moving in midair. As athletes were preparing for Olympic

---

* The 2020 Olympics took place in 2021, delayed by the the COVID-19 pandemic.

performances, I was gearing up for academic ones. After scrawling two hundred pages in an effort to finally finish my PhD dissertation, I too felt disoriented by the twists and turns of the final stretch of my doctoral program, which I was now calling Mission: Dr. Justice.

Like preparing for the Olympics, the final events of a doctoral program require years of training and qualifying rounds. Getting admitted into the graduate program is like joining the national team. Not everyone makes the cut. At MIT Media Lab the admissions rate was less than 8 percent—still much better than the odds of making the U.S. Olympic Team, which took only six elite gymnasts. Making the team is one point of celebration yet only the beginning. Once I passed my PhD general exams in August 2019, I qualified to prepare for the PhD defense and a written dissertation. My 2021 summer plan was to use May through July to write the full draft of my dissertation and then defend the work on July thirtieth. I tapped into my former years as a pole vaulter and developed a dissertation boot-camp regimen: I wrote on average five hours each weekday with Gregorian chants as my acoustic companions. Meditate. Work out. Write. Sleep. Repeat. After weeks of focused efforts, I finished the first draft.

Like Simone Biles I had made it through the preparatory stage for the finals. Then I experienced the doctoral twisties, a derailing phenomenon that occurs when a PhD student can no longer perform the maneuvers to complete any program requirements and loses her sense of direction. Staring at my laptop screen, I couldn't bring myself to make any more edits. Scrolling through my drafts filled me with dread as I realized there were even more things I wanted to add, but there was simply not enough time. I didn't want to let my parents down. I thought of my grandmother in Ghana, who was eagerly cheering me on to finish. Emotionally drained and intellectually exhausted, I was uncertain by July twenty-seventh if I could face my dissertation committee, who were set to put my knowledge to the test. I had a new set of guardians of the Algorithmic Justice League. My committee included Ethan Zuckerman, my doctoral advisor, and Hal Abelson, a distinguished MIT professor, two members from the original "Gender Shades" committee that supported my master's thesis. In addition, I was lucky enough to have Professor Catherine D'Ignazio, who had been a former student of Ethan's in the Civic Media group, and Professor Latanya Sweeney, who had finished her doctoral work under the guidance of Hal Abelson. I did not want to disappoint them either, but I also could not ignore my bone-numbing exhaustion.

Feeling defeated, I texted Ethan. "I don't have it in me anymore. I really think I should just drop out. It's too much," I lamented.

"Are you sure you want to drop out? You are so close. Everyone hates the writing process. If you really want to drop out, that is your decision, but consider postponing instead. Take some time off and think about if you really want to take that step. Postponing buys you time to make the decision; dropping out now would make continuing in the future harder."

"But all the time I'm spending on the PhD is time I am not spending on other responsibilities, and we just hired three new staff members at AJL and are currently in the process of hiring two more people. Then, I have media obligations in September. It's too much!"

"Give it some thought, and take care of yourself."

I emailed my committee and a set of mentors who had been receiving weekly updates on my progress toward Mission: Dr. Justice.

Dear Committee,

Thank you for all of your time and energy in helping me develop a strong thesis. The process has enabled me to refine concepts and also reflect on the body of work I have done during the course of my doctoral studies. AJL has been strengthened by this work. The 200 pages drafted have incubated many promising explorations including the evocative audit that will inform future work.

However, I have reached burnout mode. I was hoping to be able to make this last lap, but I am out of energy and also have to decide what I want to prioritize with limited capacity. The last 4 years have been an exercise of pushing my body beyond its limits. This is something I will no longer do for the sake of pleasing others. I still remember being in the ER the night before my oral general exams after submitting two congressional testimonies and a briefing for the European Union's Global Tech Panel for the written requirements. I left the hospital in the morning, went home to put together a slide deck, presented in the afternoon, and passed. Looking back, that is not the example I want to set. Pushing myself to the brink at all costs is not worth it. Ignoring my body is not worth it. I have 3 academic degrees: BS Computer Science, Georgia Tech (Highest Honors), MSc Learning & Technology, Oxford University (Distinction) as a Rhodes Scholar, and MS Media Arts and Science, MIT (landmark "Gender Shades" work) among other academic and professional achievements. I reject the notion that all of the past work I have done is somehow less without a 4th academic degree.

Working to finish this PhD continues to take me away from the work that having a PhD is supposed to enable. This summer when I passed on doing congressional engagements to push for critical legislation on facial recognition technologies and also declined a request from the Secretary General of the United Nations, it really made me question my priorities and where I want to have impact.

If the aim is to share concepts and work that I believe are important with the world, I have AJL, a platform, and [an upcoming] book that will enable me to do so with a much larger audience.

If the aim is an advanced MIT degree, I already have one. . . .

Given your experience, I certainly welcome your thoughts on the value of this degree specifically for me, given everything else I am doing and have put on pause.

Unfortunately, or perhaps fortunately for my body, I will not be defending on July 30th as initially planned. I apologize for the inconvenience this will cause and this is not a decision I make lightly. I've been waiting a few days to see if my body would rally. It has not.

With that message sent, I gave myself a month off from my official work as a PhD candidate and my de facto work as the executive director of the Algorithmic Justice League. Simone Biles's example of self-love and self-preservation could not have come at a better time. I wrote her a public letter of gratitude:

Dear Simone Biles,

Generating **O**utstanding **A**wareness **T**enaciously! This week the world witnessed a great skill: the Biles Refusal—executed by saying no to golden promises to say yes to priceless health—delivered with grace. Thank you for your life affirming example. You are courage personified.

Your actions have inspired me to set boundaries I thought were not possible because of the weight of expectations. Too often I have sacrificed my health for seemingly golden achievements and implicitly tied my worth to high performance. Too often I have put my needs last to please other people. Too often I have said no to my own happiness as if it were some noble sacrifice to be a martyr for a cause. Too often I have felt obligated to achieve even more than I already have in order to prove those who doubt my intelligence and worth wrong. Too often I have committed to near impossible workloads, because I can.

The Biles Refusal is a beautiful reminder of the power of saying yes to your well-being.

It is not a badge of honor to be burnt out. It is not a sign of fortitude to over commit. Putting yourself above the weight of gold is awe inspiring for me to see especially as a young Black woman. We know that when we say no, it disrupts expectations that we should be meek and grateful to have opportunities to shine on the world stage. We owe ourselves the compassion and care rarely extended to us by the same people who marvel at our ability to endure injustice.

Despite your undisputed greatness as the most decorated gymnast of all time made possible with your pairing of immense talent and diligent practice, you do not face a fair playing field. I cringe when your skills are not given the scores they should command because judges claim they want to dissuade those with less ability from executing them. Capping someone's greatness and changing the rules when someone succeeds shows a glaring truth. When individuals framed by society as inherently unworthy based on their gender, race, and background succeed anyway and outshine the competition, the establishment seeks to restore the old order. The old order is crumbling because after enduring a pandemic, continuously experiencing injustice no matter the level of success, and embracing their inherent dignity no accolades required, queens like you are embracing refusal.

Thank you for demonstrating a better order where health comes before gold, where self-preservation comes before thankless martyrdom, where the dreams of daughters of diasporas to be loved for who they are, not what they do, come true.

With love and respect,
—Poet of Code

Coming to a halt after such an intense period of work was not easy. Since I had shared the concept of the Algorithmic Justice League at TEDxBeaconStreet five years earlier, in 2016, the organization had grown to five full-time staff. I was still the organization's number one volunteer and not on the payroll. I was eager to grow our capabilities, as the fight for algorithmic justice needed as many helping hands as possible. I was also eager to not have a PhD looming over my head so I could start a full-time role at AJL. One of my MIT professors had joined to lead AJL research so we could continue to provide evidence about AI bias and harmful use of AI systems. A week before taking time away to work on my dissertation I had brought a fundraiser aboard so I could spread the responsibility of securing money for the organization. Before they joined, I had been fortunate enough to secure millions of dollars for the organization.

The Ford Foundation, the MacArthur Foundation, the Sloan Foundation, and the Rockefeller Foundation took a chance on funding AJL, a small organization led by a kid who had yet to finish graduate school. Balancing the expectations of a PhD program and leading a growing organization left me perpetually drained. Between reading research papers, I would have meetings with program officers, sort through the dozens of weekly requests, and do my best to track policy developments in the United States and abroad, while giving unending media interviews and conducting weekly team meetings and one-on-ones with staff. We were looking to hire two more people to spread the load. Once I decided to take a break full stop, I had to update AJL staff and make alternative hiring plans. It took about a week to actually slow down. I interviewed a finalist candidate for a communications position, convincing no one that I would actually take time off. I didn't want to halt all the momentum that was building around the organization, but I needed a moment to breathe.

Free time felt strange. There was nothing I absolutely had to do, so I started to focus on what I wanted to do. As I continued watching the Olympics, a commercial came on that showed spirited street athletes from my homeland of Ghana. The 2020 Olympics were special not only because the competition was taking place in 2021 due to the pandemic, but also because it was the first time skateboarding, my childhood sport of choice, was part of the Olympic Games. Skaters like Japanese British Sky Brown had become common poster kids for the sport, but the Ghana skate commercial was the first time I was seeing skateboarders who looked like me in the streets of my homeland being celebrated.

Instead of being an academic, maybe it was not too late to become a professional skateboarder and maybe even represent Ghana at the next Olympics. Paris 2024 was only a few years away, giving me enough time to improve my dormant skills. What if I dropped out of MIT and became an Olympic skateboarder? I watched the street and vert competitions and saw women of all ages, from teens to those in their thirties, showcasing the sport. One of the skaters on the U.S. women's street team had earned her master's in mechanical engineering from MIT. Soon I convinced a friend to join me in my fantasy. We went to a Vans store at a local mall, where I purchased red and white waffle-heeled skate shoes. They perfectly matched the new Dr. Justice Skateboard I had had custom made. The black deck had "Never Stop Dreaming" in pixelated white font on the nose of the board. The top of the board was covered with red grip tape emblazoned with the stamp "Algorithmic Justice League." Well aware that my bones were not quite what they used to be, I invested in protective gear. I finished off my white helmet with a silver rainbow decal that said "Poet of Code" in graffiti-style lettering. With a pink camouflage-print neck gaiter, I headed off to a nearby skate park on a sunny summer Tuesday.

The skate park was situated next to a blue and green asphalt basketball court. The concrete maze had thirty-six-inch drop-ins and flanked the left side of the basketball court. I sat in the grass observing other concrete surfers. Eventually I introduced myself to a lanky man with shaggy dirty-blond hair. Before long the topic of age came up. Most of the people at the park that day were in their thirties. My first thought was *What are these thirty-somethings doing at a skate park on a random weekday at one in the afternoon?* Then I realized I was among their ranks. I stopped asking questions and finally put my Dr. Justice board to work. As I skated, a mom and her daughter stopped by and smiled at me as if to say "Go, girl!" For the next three weeks I found solace in imagining myself as an Olympic hopeful preparing for the Summer 2024 Games. The academic games faded for a while. The sound of ice cream trucks replaced the Gregorian chants that permeated the first half of my summer.

Later that August, news of a Taliban takeover in Afghanistan filled international headlines. I saw stories of Afghan women hiding their diplomas and worrying that the takeover would mean the opportunities afforded by their educational attainments would no longer be possible. The news of war jolted me out of my skateboard escape. If I returned I would be a few months away from completing a PhD from MIT. In the United States less than a percentage of a point of computer science–related PhDs went to Black women out of only 55,283 people who received research doctorate degrees from U.S. institutions in 2020.[1] Seeing women who would no longer have the choice to make academic achievements made me appreciate how much I was taking the right to education for granted. I also thought about the history of MIT. In its early years, enslaved people worked in laboratories, libraries, classrooms, and gardens.[2] In the nineteenth century, were I enslaved, it would have been illegal for me to be taught to read or write in the state of Georgia, where I earned my undergraduate degree. I thought about visiting the National Civil Rights Museum in Memphis, Tennessee, and going through exhibitions about freedom fighters who fought against Jim Crow segregation and bled in the streets and languished in jail cells so people like me could have the option to study at educational institutions. I thought of James Meredith, who in the 1960s attended the University of Mississippi, risking his life to test the bounds of the law. I thought of visiting my father's chemistry lab on the Ole Miss campus three decades later. I thought of my grandfather, who earned his doctorate in medicinal chemistry at Chelsea University in the United Kingdom in 1969, less than a decade before women were allowed to study at Oxford University as Rhodes Scholars in 1976. During the celebration marking the fortieth year of women Rhodes Scholars, Secretary of State Hillary Clinton in her recorded remarks stated that when she was at the age where young men like her husband could apply, she could not. It was no mean task that I had been one of three Black women from the U.S. set of Rhodes Scholars in 2013 to attend and graduate with distinction from the fabled dreaming towers. It was no easy task to earn a master's degree from MIT two years later.

I thought of Dr. Latanya Sweeney, who now served on my committee and had been the first Black woman to receive a PhD in computer science from MIT in 2001. She encouraged me to continue. "Joy, the opposition is getting ready to push back on all the progress made in the movement for algorithmic justice; you don't want to give them a foothold for attack. The choice is yours. I can tell you despite how hard it was for me to get mine, having the PhD from MIT has opened opportunities I otherwise would not qualify for and made it harder to dismiss my presence. When you are advising, your trajectory as a PhD candidate from MIT has added to your credibility."

Dropping out of MIT in the long run might not impact me significantly in terms of job opportunities, as many thrive in the tech field and other spaces without graduate degrees, let alone doctoral degrees. Still, the stereotypes associated with dropping out as a young woman in STEM felt unfair given all the work I had already done. My aim was to do a poetic PhD and with "AI, Ain't I A Woman?" I had shown myself what it meant to be a poet of code. With influential research publications, I had proven I could make meaningful academic contributions.

I was annoyed by the notion that not being validated by an academic institution would be a potential scarlet letter. Perhaps I should have just concluded pursuing academic degrees after earning my third one in 2017. I wondered if this was just a minuscule taste of how Serena Williams felt when people pressed her about obtaining a twenty-fourth grand slam championship, as if her twenty-three championships did not already cement her powerful tennis legacy. I also thought about the professors and faculty at MIT who did not have PhDs yet continued to make significant impact. During my time at the Media Lab, the head of the academic program did not have a PhD. I also noticed that at the time I was making these deliberations all the women on the Media Lab faculty had PhDs, and many were credited with starting new fields of research, including Dr. Cynthia Breazeal, who pioneered social robotics, and Dr. Rosalind Picard, who created the field of affective computing. Most recently, Dr. Danielle Wood had established the Space Enabled group, looking at the possibilities of space technology and social justice. In other spaces, being a dropout was an iconoclastic badge of honor. Dr. Wood attended MIT, where she earned a PhD in engineering systems, an SM in aeronautics and astronautics, an SM in technology policy, and an SB in aerospace engineering. Famous dropouts like Steve Jobs, Bill Gates, Richard Branson, and Elon Musk were often celebrated in the press. Larry Page and Sergey Brin, the founders of Google, took a leave of absence from their PhDs at Stanford and never returned. What was different about my case?

The ironic stigma of being a dropout hung in my mind, but more important, the legacies of so many people who made this choice possible permeated my thoughts. What did I owe the past? What did I owe the future? What did I owe myself? Dr. Sweeney was now paving the road for me again, reminding me of what was at stake if I chose to drop out. August had been lovely, short, and full of sleep. But I had decisions to make, and September promises to keep.

CHAPTER 20

GOLDEN REDEMPTION

Vogue September 2021, the most coveted print space in fashion, sat on top of my Dr. Justice skateboard. I hurriedly thumbed the pages of the magazine until I found the multipage spread for Olay. I saw a familiar face: mine, frozen in a sly smile—the face of a multimillion-dollar national ad campaign. My skin gleamed in a red sleeveless dress that showcased my pole vaulter arms and hugged my bootcamp-chiseled figure. My red glasses balanced a golden wrist cuff with sumptuous curves. My natural hair was adorned with galaxies of golden spheres that matched my gold-chromed fingernails. In the background were faded glimpses of the kind of women you would typically see in the foreground of beauty campaigns, slight milk-skinned figures with straight hair. On top of the image sat the motivating question for the campaign: "Where are the women of color?" This *Vogue* ad was the exception to the rule; the representation of beauty overall was heavily Eurocentric, and search engine results for terms like "beautiful skin," "beautiful face," and "beautiful woman" revealed and amplified this beauty bias. At that time those search terms brought up predominantly light-skinned women.

If you had told nine-year-old me that one day I would be the face of any beauty campaign that celebrated my dark skin and Ashanti features especially, I would not have believed you. I was still nursing stings that would continue into adulthood of being shaded for my complexion. I remember the schoolchildren who would put their arms next to mine and sigh in relief that their skin was not as dark. If you told me the beauty campaign would somehow be linked to science, technology, engineering, and mathematics (STEM), I would be further confused. Yet in 2021 I became the face of Olay's #DecodeTheBias campaign and in the process worked to increase the number of women in STEM; develop guidelines for creating just, responsible, and inclusive consumer AI products; and elevate public awareness about issues of algorithmic bias. I still remember my spokesperson talking points.

The ability to use a beauty campaign to educate more people about the shortcomings of technology was my main motivation to work with Procter & Gamble, one of the world's largest advertisers. The opportunity to work with Madonna Badger, a legend of empowerment advertising that celebrated women as more than objects, also drew my interest. When I was invited to consider a STEM ad campaign with the company, I did an overview of prior P&G ad campaigns. Their Emmy-nominated "The Talk" and "The Look" campaigns both addressed racial bias. P&G furthered the conversation with the #TalkAboutBias campaign that provided a discussion guide, organizations to support, and actions to take to address racial bias in society. "The Talk" (2017) and "The Look" (2019) were released as the cry for Black Lives Matter was gaining stronger mainstream awareness. In 2020 P&G released "The Choice" and explicitly called for a need to be anti-racist and take action to combat racism, following the murder of George Floyd. P&G as a company increasingly made moves to use their advertising to advance difficult conversations. This is where I saw the opportunity. I even sent P&G an email describing how I thought the company could use its platform to open up a national conversation about how machines can give a look, a gaze that produces harm by reducing individuals to stereotyped labels.

"The Look," like the coded gaze, is a reflection of societal cues of who was deemed suspicious, who was assumed to have authority, and whom our children were taught to shun. The ad follows a day in the life of a Black man who is facing everyday yet no less painful discrimination as he is followed in a department store as if he could be a thief and distanced by workplace colleagues as an unwelcome presence. The final scene reveals he is in fact a judge, but even that standing in society when he leaves the bench of authority is no shield for microaggressions. The scene reminds me of Bryan Stevenson, executive director of the Equal Justice Initiative, who describes the experience of being arrested while sitting in his own car and having to keep the wherewithal to calm the police, or being mistaken for the defendant facing discrimination in courts as a Black lawyer before even starting the work. I remember the many times I walked into rooms, even ones where I was invited to speak, and was treated as if I was taking up space meant for important people. Afterward, apologetic organizers would reassure me that these incidents did not reflect their views of me. Still, the daily slights take their toll.

And then there are the more egregious personal attacks on

dignity. After I was in touch with *60 Minutes* producers for months on a story they were doing that featured facial recognition technologies, they scheduled me to do an on-camera interview with news anchor Anderson Cooper. For the interview, I had spent many late nights creating a custom-built interactive demo to show the different ways facial recognition technologies could fail and preparing my talking points about algorithmic bias based on my research. I gathered demo images of Anderson Cooper as a young reporter cued up to personalize the examples I had planned. I thought about short sound bites to say during the interview to make it easier for everyday people to understand why the demonstrations mattered. I spent considerable time with the producers, answering technical questions and pointing them to additional experts to interview, including Patrick Grother of NIST, who had extensive history with assessing the performance of facial recognition algorithms through the U.S. government. From my perspective, he was definitely someone to talk to and interview.

In February 2021, on my way to get a COVID-19 test for the interview with Anderson Cooper, the producers messaged me to say the interview was canceled. I felt the anger rising inside me as I glanced at my cellphone and reread the message. All the thought and care I had put into preparation felt wasted. My heavy disappointment sat with me as I told the driver to take me back home. Back at my apartment, I looked over at my backpack, stuffed with two silver laptops I had configured for demonstrations the next day. Well, even if I wasn't interviewed on camera, I reasoned, at least they would share the "Gender Shades" research.

When the segment aired in May, Patrick Grother offered his perspective on the current state of facial recognition. The segment touched very little on the evidence NIST itself had collected about algorithmic bias on the basis of race, gender, and age. There was no mention of any of the research I had published with Dr. Timnit Gebru and Deb Raji. I felt both exploited and erased. Worst of all, the viewers were getting an incomplete picture of the varied state of a range of facial recognition technologies. At first I thought I would just ignore the situation, as the episode aired on the day I began drafting my PhD dissertation. The thought crossed my mind that perhaps not having a PhD justified my exclusion, but then again Dr. Timnit Gebru was not included either. Clare Garvie, a brilliant legal expert without a doctoral degree, was interviewed on camera. She and I had both been witnesses at the 2019 congressional hearing on facial recognition technology. I could not blame a lack of credentials.

Though I was tired, if I did not speak up for myself I would be quietly complicit in this kind of treatment. Though I had only eight weeks until my PhD defense, instead of working on my long-deserted dissertation, I reached out to the producers to ask about my exclusion. They had not been in touch since the February cancellation. They told me simply that they don't include everyone they interview. I could have left it at that, but I had given them so much of my time for free that it felt wrong to act as if nothing had happened. I told my parents about the *60 Minutes* incident and asked their opinion. "You need to focus on your PhD. Keep your life simple. You don't have to fight everything," my mom cautioned. I explained the situation to my dad. "They did what? Kosiesem. Take them to task!" he encouraged me. I'm a daddy's girl. I spent the next few days working with several teams on a public petition that demanded CBS take corrective actions.

The petition received more than six thousand signatures. CBS released a statement:

In response to our story "Facial Recognition" (airdate May 16, 2021)—about law enforcement's use of facial recognition technology to identify suspects—we heard from some viewers who believe we should have included the work of computer scientist Joy Buolamwini and the organization she founded, the Algorithmic Justice League, regarding algorithmic bias. . . . We are very grateful to the dozens of sources— off and on camera—who helped us develop and focus this segment but were not mentioned by name. As with all our reporting, we spoke with a wide range of people, including some of the leading thinkers and researchers in the field, like Ms. Buolamwini.[*]

This statement made me wonder who else had given significant uncompensated time to inform the segment. Even if this was common practice, normalizing erasure and the glib use of behind-the-scenes labor was not something I could sit by quietly and witness. The additional labor of then having to combat erasure while needing to work on my PhD added salt to the wound. Time spent on the work I had been postponing in an effort to educate the public about algorithmic bias was now being delayed to educate the public about erasure. The work before the work after doing the work was heavy. Being erased by machines had become a familiar story to me from a personal standpoint, but this media erasure had a much longer shadow.

When I shared my erasure experience on social media, many people came to my defense, and I was introduced to the work of Christen A. Smith, who started the #citeBlackWomen campaign. Alvaro Bedoya was one of many allies on social media. While his

* You can read the full statement at www.cbsnews.com/news/facial-recognition-editor-note/.

words were encouraging, I nonetheless felt discouraged. If after all the work I had done since 2016 to specifically bring public awareness to algorithmic bias I could be so easily sidelined, why keep putting in so much effort toward engaging with the media? I also gained a newfound appreciation for the power of the *Coded Bias* documentary then streaming on Netflix. *Coded Bias* centered so many women and especially women of color as experts on the social implications of AI and algorithmic bias. In the documentary Shalini Kantayya was combating symbolic annihilation through her selection of subjects to follow. Her tenacity in navigating the documentary world made it possible for the origin story of the Algorithmic Justice League to reach so many people.

My ego was also bruised. I had gotten used to the spotlight and being among the leading voices when it came to issues around facial recognition technologies. *Fast Company*'s choice to put me on the cover of the "2020 Most Creative People in Business" issue under editor Stephanie Mehta was, I felt, a significant step in media representation, given how few Black people who were not professional athletes or entertainers were allowed such a spotlight. Looking at it also made me feel validated as I stared defiantly in a yellow jacket. Soledad O'Brien's 2018 visit to the Media Lab to interview me about my research now took on greater meaning. I could not take it for granted that other producers would take the time and care that I had seen from her team. These journalists used their decision-making power to highlight meaningful research on algorithmic bias and give me access to the platforms they controlled. Comedians like Trevor Noah and John Oliver, who used humor to talk about racial bias in AI, using the white mask example and the "Gender Shades" research, also increased the reach of the work. They took the time to explain the implications of the findings while crediting the source. Although I had previous impactful representational opportunities, the *60 Minutes* case was a sobering reminder of the work to be done. It also made me question my relevance. Maybe I was old news. I would have to continue to find gate openers to counteract the gatekeepers of mass media. I would also work on being a gate opener myself. The removal of the spotlight burns.

Besides facing erasure from machines, I have witnessed the symbolic annihilation of women of color and dark-skinned women. Symbolic annihilation describes the absence or lack of representation of a particular group. In journalism, whose work is highlighted sends a message about who is viewed as an authority and the face of expertise. If news coverage depicts Black people only as victims, it perpetuates a harmful trope that we lack agency to make meaningful change. In advertising and marketing, who is represented is one part of the story in terms of what society decides to celebrate. Complete erasure is one way to "invisibilize" a group, yet inclusion that builds on stereotypical representation can also be harmful. It is not enough just to be seen if you and people like you are rendered in disempowering terms or through disempowering frames. By continuing to show up even when it hurts and even at the risk of erasure, I was combating symbolic annihilation. Now I knew I would need to be more strategic about where I committed my energy.

As I was rekindling my energy, my longtime collaborator Dr. Sasha Costanza-Chock sent me a video message while at the local CVS. Holding up a *Cosmopolitan* magazine, they shouted excitedly, "Look what we found!" In addition to running print campaigns with *Vogue,* Olay also ran ads in many major beauty magazines including not just *Cosmopolitan* but also *Allure* and *Harper's Bazaar.* This was phase one, and it was now time for phase two of #DecodeTheBias, which would include TV commercials, a satellite media tour, and a segment on *Good Morning America* on ABC. The kid who wasn't allowed to watch commercials was now starring in one. The door to reaching a national TV audience that had been closed by *60 Minutes* was being held open by Olay.

CHAPTER 21

COSTS OF INCLUSION AND EXCLUSION

I n the lead-up to the satellite media tour and the filming of the *Good Morning America* segment, I worked with the Olay team on the materials for the press release and campaign website. Freshly reminded of the sting of erasure, I started to think how I might use this media spotlight to highlight the important work of others. To the press release I added a reference to the work of Dr. Safiya Noble, author of *Algorithms of Oppression*, whose path-breaking work on racial and gender bias in search engines power-fully documented how Black girls and women of color were portrayed in derogatory ways. Searches for "Black girls" or "Asian girls" pulled up hypersexualized and at times pornographic im-ages, while searches for white girls did not have similar top re-sults. The resource section included her book along with work from leading femme thinkers, including Dr. Ruha Benjamin's *Race After Technology* and Dr. Sasha Costanza-Chock's *Design*

Justice—the book that opened me up to a greater understanding of the harms and erasure often experienced by trans folks and broadened my thinking on what it meant to reach people of many genders with a beauty campaign. In addition to elevating the work of leading thinkers, I also suggested that Olay include Black Girls CODE to support their aim of tripling the number of women of color in STEM. They agreed! The call to action from the campaign would be to use the hashtag #DecodeTheBias on social media to send girls to code camp. This work also supported AJL's strategic priority of elevating the excoded, those who are harmed by AI systems. I was not sure how much I would be able to influence the campaign, and I was wary of the possibility that I would be obligated to deliver lines I did not believe in. I had my doubts.

When Olay first approached me, I took their proposal to the AJL board members for their response. Overall, the board members saw the potential for reach, but we also agreed that the framing would need to change. We got into conversations on just how much can be conveyed in a sixty-second ad and, given the constraints, what was most important to highlight. I was surprised when Olay and Madonna Badger brought me on as a creative partner and collaborator. They sent me documents outlining the design concepts for the commercial and invited my poetic voice into the script. One of the first versions of the campaign was centered on the idea of inclusion and collecting face photos that might help improve the performance of Olay's Skin Advisor system. After some back-and-forth about how being coded in without having explicit agency might actually set people up for surveillance, we deliberated the question: "Code me into what?"

Still, the Olay team pointed out that in my TED-featured talk I had called for a "Selfies for Inclusion" campaign as a way of addressing algorithmic bias. They were right: I had offered what appeared to me to be the most straightforward action to address algorithmic bias: creating more inclusive datasets. I had given in to well-meaning pressure to end my talk with something people could do, so that I was not just the messenger of doom and gloom. At that stage in my development I believed in a penchant for action. To an engineer action meant building something; thus, my default when faced with a technical challenge was to come up with a technical solution: build a better dataset. However, when looking at the deployment of facial recognition technologies into the real world we are looking at a sociotechnical problem, which was a departure from my computer science education. With a sociotechnical problem, code and data are not enough, because the issues extend beyond just how well a given system performs into, even more important, how a system will be used. Who will benefit? Who will be harmed? Who gets to decide? These were questions of power, not questions of performance metrics. In 2016 I reflected on my TEDxBeaconStreet talk in an article where I wrestled with the tension of different approaches to addressing algorithmic bias:

> Alongside tools to rigorously identify bias, there also need to be ways to mitigate bias. Mitigating bias is not just a technical challenge. How and when machine learning should be used is a matter of ongoing discussion. Questions of appropriate mitigation approaches remain.

The Costs of Inclusion
If bias is identified, should we stop using the software altogether or work towards minimizing bias? If AJL launched a #selfiesforinclusion campaign that improved facial rec-

ognition, are we inadvertently subjecting more vulnerable
populations to unfair scrutiny? How can such risks be addressed? Who should address them?

The Costs of Exclusion

If we do not improve the systems and they continue to be
used, what are the implications of having innocent people
identified as criminal suspects? Considering the advent of
self-driving cars, can we afford to have pedestrian detection systems that fail to consistently detect a particular
portion of the population? Who has a voice in deciding
how we move forward?[1]

As much as the engineer in me wanted straightforward answers and straightforward technical solutions, the reality is far
more complex. There are costs of inclusion and costs of exclusion to be considered in the design and deployment of AI systems
that must be contextualized. Look at the cost of exclusion. A
blanket rejection of AI systems that analyze humans in some way
would foreclose beneficial innovations. For example, AI systems
are increasingly being used in medical diagnostics. Dr. Regina
Barzilay of MIT, after surviving breast cancer, used her expertise
and personal experience to develop AI systems that have shown
deep promise in detecting early signs of the disease.[2] At a Black in
AI workshop, I was encouraged to hear a group of researchers at
the healthcare start-up Ubenwa using machine learning to analyze the cries of babies in Nigeria to detect lung conditions. Yet
we also have substantial evidence of how AI systems created with
benign intent can show biased outcomes in the medical context.
In 2017, Stanford researchers released a paper showing promising
results for skin cancer diagnosis with computer vision. They were

able to match the performance of dermatologists, and enthusiastic headlines followed. However, the dataset that was used to benchmark the performance was later revealed to be overwhelmingly lighter-skinned individuals. Thus far, this analysis is a rather technical view of the problem.

A sociotechnical view requires we think not only of datasets but also of the social conditions that led to a privileging of white skin in dermatology and how medical apartheid manifests. I was introduced to the term medical apartheid while serving as a judge for a Mozilla competition. Avery Smith was one of the applicants who went on to win a grant. He wrote powerfully about losing his wife to melanoma. By the time she was diagnosed, the cancer had progressed beyond repair, and her story fit the pattern that dark-skinned people are often diagnosed with skin cancer that has entered much later stages. Medical training is one part of the problem. Many medical textbooks and examinations use light-skinned representation of dermatological issues so that by the time medical students become clinicians they have been exposed to very limited representations of the disease. Another issue is gender and racial bias in clinical settings. The dismissal of patient concerns about their health, especially the pain of patients of color, women, and women of color also means that even if we do go to the hospital for help, our symptoms and pain can be discounted or underestimated. Other parts of the problem are the lack of dermatologists whom patients feel they can connect with, and justifiable distrust of the healthcare system given historic injustices.* To address these issues Avery launched Melalogic, which addresses social and technical aspects of the

* Consider, for example, the Tuskegee Syphilis Study (www.cdc.gov/tuskegee/timeline.htm), and the forced sterilization of women in Puerto Rico (www.library.wisc.edu/gwslibrarian/bibliographies/sterilization/).

problem of dark-skinned patients not getting needed dermatological care in a timely fashion. On the social side the company
was creating a database of Black dermatologists to make it easier
for patients to find someone they felt they could relate to. On the
technical side it was creating a dataset to gather more examples
of dermatological conditions on darker skin. Avery's work, motivated by deep personal loss, points the way forward in creating
future AI systems that attend to social, cultural, and historic realities.

On the other side of the equation are the costs of inclusion.
While more diverse dermatology datasets can help reduce medical apartheid, inclusion can have its downsides. Because my research on algorithmic bias was motivated by facial recognition
technologies (FRTs), I immediately had to contend with the
harms of inclusion. Diversifying datasets with the aim of improving the performance of FRTs in a vacuum might seem like a
straightforward fix to algorithmic bias. Yet algorithmic bias—
when a system performs better on one group compared to another—is only part of the conversation. When we think about
algorithmic justice—shifting power so the burdens of technology
do not fall on the marginalized and the benefits do not accrue
only to the privileged few—we have to think about algorithmic
harm.

Even if FRTs were technically flawless, more accurate systems
could still be abused. They can be enlisted to create a camera-
ready surveillance state. They can be developed to recognize not
only an individual's unique biometric signature but also soft biometrics like age. Technology that can somewhat accurately determine demographic or phenotypic attributes can be used to profile
individuals, leaving certain groups more vulnerable to ill-justified
police stops. An investigation by *The Intercept* reported that IBM

used secret surveillance footage from the New York Police Department and equipped the department with tools to search for people in video by hair color, skin tone, and facial hair.[3] Such capabilities raise concerns about the automation of racial profiling by police. Police brutality, which can be furthered with the strengthening of surveillance technologies like facial recognition, which itself can undermine civil rights and liberties, is a life-taking and life-threatening harm that extends beyond technical algorithmic bias. The consideration of expanded algorithmic harms, instead of narrow algorithmic bias, is necessary to address what the nonprofit think tank Data & Society has termed the "specification dilemma." For example, where harmful impacts are narrowly defined to focus on technical performance of facial recognition systems, people who would be harmed from mis-identification can be accounted for, but the impact of mass surveillance falls outside this understanding. As I wrote for *The New York Times,*

> It's important to remember that even if false-positive match rates improve, unfair use of facial recognition technology cannot be fixed with a software patch. Even accurate facial recognition can be used in disturbing ways. The Baltimore police department used face recognition technology to identify and arrest people who attended the 2015 protests against police misconduct that followed Freddie Gray's death in Baltimore. We need to challenge the growing use of this technology.[4]

So by the time I was having a conversation about potentially collecting images for the Olay campaign, I had greater clarity. I had much more experience understanding the harms of facial recognition technologies that resulted from more than performance issues. I also had more tools to use than my engineering skills. I had testified for a federal moratorium on FRTs. I had supported successful resistance campaigns to harmful use of FRTs. And I had nixed the launch of a face data labeling project I was working on for months. I had also faced quite a bit of criticism for the TEDxBeacon Street call to action. By moving to send girls to code camp instead of collecting selfie images, the final call to action looked at not just how technology was being created but who was creating technology. Sending girls to code camp was a step in addressing structural issues that led to algorithmic bias and beauty bias. To address beauty bias we needed more "women leading and coding in their beautiful ideas."[*]

We also needed more companies stepping up to check their AI products with algorithmic audits. The final component of the #DecodeTheBias campaign was an audit of Olay's Skin Advisor tool. When I was first asked to audit the tool, I warned the team that given everything I had learned about how it was developed there was a high chance we would find unflattering bias. I also did not want to be part of conducting an audit where the final results could not be made public. The Olay team did not back down. "If we find bias in our tool, we will do what it takes to fix it."

I asked, "What if it cannot be fixed?"

Olay's response was: "If there is no way to make it more just and inclusive we will shut it down." I was shocked.

The option to not proceed or to stop building a system was rarely on the table. "All right, if you really want to do the audit and we agree in writing that we can publish the results, I can help

---

[*] This line is a phrase I added to the #DecodeTheBias commercial script.

you with all pieces of the campaign." After the meeting with Olay, I called my blue-haired fellow freedom fighter Cathy O'Neil, the founder of algorithmic auditing company ORCAA. "What do you know about skin care?"

"Basically nothing."

"No worries, I have an epic audit we should do."

"Tell me more."

The ORCAA audit revealed that the Skin Advisor app did indeed have algorithmic bias on the basis of skin color and age. The company published our findings on the companion website dedicated to the #DecodeTheBias campaign. Crucially, I had final editorial control of the report that came out, which meant that even unflattering results were shared. P&G went a step further by committing to take actions that addressed our recommendation with time-bound dates. To me the most important action they took was agreeing to the Consented Data Promise, a commitment to use only data collected with explicit user agreement. I came up with the name by taking inspiration from the Olay Skin Promise (zero skin retouching in all ads). By the time I was the face of this campaign, the company was no longer airbrushing models or spokespeople. The images and videos in the campaign represented what was captured without post-production blemish-reducing techniques. While I appreciated the intent of the Skin Promise, a part of me was secretly disappointed. I wanted the assurance that if I should have an ill-timed pimple, the airbrush would save me. Knowing I had no backup, I started what I called "beauty bootcamp." I did all the right things for vanity reasons. I drank mainly water and removed all caffeinated drinks from my diet. I worked out five times a week and went to bed at a reasonable hour, averaging eight hours of sleep a night. And, of course,

I used the Olay skin products that adorned my sink. Eye creams, face creams, serums, clay masks, and a mist that smelled of cucumbers.

Just like knowing I could not hide behind post-production retouches, having companies undergo public-facing audits can apply positive pressure. While Olay sought the audit on a voluntary basis, we cannot assume all companies would do the same. While there may be some good actors, relying on the goodwill and moral motivations of the tech industry is not a responsible or reliable strategy. The year after the Olay campaign, I would also see firsthand that government agencies would need to be pressured to test technology from companies offering AI services. For now, with beauty bootcamp complete, it was time to return to my PhD. The time off I'd taken had costs. I had hoped those looking at the magazine ads for the campaign would see "Dr. Joy Buolamwini, Founder of the Algorithmic Justice League and Poet of Code." Missing my defense date meant I had not yet earned the honorific. Reaching out to the PR firm for the campaign and telling them to remove the "Dr." title from the press release and print campaigns was the most deflating part about the process.

When the campaign did launch, I was fortunate to have a glam squad assigned to prepare me for the satellite media tour. Amy, a tall Nigerian woman, was my makeup artist. When I showed her the print ad, she looked at me with tears welling in her eyes. "I know this is supposed to be for little girls, but it's for me too. We don't get to see ourselves celebrated like this." Her reaction impressed on me the joy of inclusion. I didn't need another credential to have others feel seen or to be taken seriously.

CHAPTER 22

SWORD OF KNOWLEDGE

I n addition to my summer 2021 burnout, a developing institutional crisis complicated my academic path. In 2019, the Jeffrey Epstein scandal had broken. Reports revealed he had contributed money to multiple research labs at institutions including Harvard and MIT. MIT Media Lab faced the brunt of public criticism. When Ethan learned that Joi Ito, the lab director at the time, had actively solicited funds from Epstein despite knowing his affiliation with sex trafficking, Ethan made a moral stand that impacted all the staff and students in our research group. He decided he could not in good conscience remain part of the Media Lab and decided to close down the Center for Civic Media. I supported his decision, but I was also left uncertain about how to continue. Just as AI systems can determine life opportunities, decision-makers made choices outside of my control and with no transparency, dramatically shifting my reality. At the time, Ethan still had a number of graduate students who had not finished our degrees, so we had to find creative ways to navigate the circumstances. For the 2020 spring semester, I had escaped the lab and gone to Emory University to continue my work in Atlanta, away from the scandal. In my personal life, I thought nothing could overshadow the events unfolding at the Media Lab, until the COVID-19 pandemic swept the globe. Working on a degree with an advisor who was no longer at my host institution, settling into a new university while navigating lockdown, and fundraising for a small nonprofit was a crucible like none other. Quitting was becoming increasingly enticing. Given the circumstances, I felt justified if I ultimately decided to leave indefinitely.

But, on October 6, 2021, fresh from the Olay satellite media tour, I was fully charged to complete Mission: Dr. Justice. Shield in hand, I strode next to Ethan into the same room where I had presented "Unmasking Algorithmic Bias" on Crit Day five years earlier. My committee members and invited guests came dressed in variations of red, gold, and black to celebrate the occasion. Ethan's six-foot-plus frame was magnificently cloaked in a maroon blazer, signaling a special occasion. I was happy to see him walking with assurance and confidence. Adhering to my dress code, I had opted for a professorial gray and black checkered blazer and a knitted gold blouse. I walked up to the podium and hung my shield on the front. Today was defense day.

My PhD sisters Alexis, who had defended during the summer, and Jaleesa, who was going through the doctoral paces, helped me to get my final items into place. Alexis made sure I had water on hand. My parents joined via video chat, and Jaleesa was double-checking the connection. They were the only people to attend virtually. Everyone else jumped through the hoops of negative COVID-19 tests to make it to the occasion. When I glanced over, I saw the name "John Buolamwini" on the computer screen. Minutes later the screen said, "John & Patricia," and I saw my parents, eager and proud, awaiting my presentation. My mother, Frema the Akan, had peppered me with so many probing questions in the week leading up to the defense that I was relieved she would not be part of the drill squad officially scrutinizing my dissertation. Shortly before the defense I did a practice run with Timnit, Sasha, and Catherine D'Ignazio, who now entered with a flowing red top to take her seat as one of the guardians of the Algorithmic Justice League.

Scanning the audience, I saw many familiar faces that reminded me of the journey I had taken through the Media Lab to arrive at this culminating hour or two, depending on how long deliberations went. Joost Bonsen, forever wearing shorts regardless of the weather or occasion, settled in. It was in his science fabrication class I had conceived the Aspire Mirror that led me to the coded gaze. My faithful friend Dr. Aubry Threlkeld, a fellow resident tutor from Adams House and now a dean, bounded in. He had sat through many hours of scheming and planning, oftentimes reminding me, "A PhD is an exercise in humility. Smile and nod and get it done." My former roommate Alicia Chong-Rodriguez, who had heard me threaten to drop out of MIT for years when I became overwhelmed, gave me a knowing look as if to say I told you so. Kevin (now *Dr.*) Hu had met with me throughout the summer to talk through my evolving dissertation. These conversations were an important part of my preparation process. They took me back to the golden age of the Media Lab when we would brainstorm ideas on whiteboards and dry-erase white walls. Every so often someone would accidentally write something on a regular wall, but it would soon be cleaned, keeping our temple of knowledge as manicured as I had always remembered it. He gave me a thumbs-up from the audience. My friend Olu-makinde Ogunnaike entered the room, with his signature curls. We had explored the possibility that I would borrow a sword from his collection to add to the symbolism of the day. When we made eye contact I lifted up my pen and moved it like a sword. I would defend my work with words. The presence of my family and close friends gave me courage.

I began the defense of "Facing the Coded Gaze with Evocative Audits and Algorithmic Audits." I walked the congregation through the history of community self-surveys and social science audits that looked into discrimination. I stated that the work of scholars like Dr. Latanya Sweeney and Dr. Safiya Noble modeled the use of lived experience to inform scholarship, and I made clear that their work had shaped my current work. I shared the debt I owed to Black women intellectuals like bell hooks, who conceived the oppositional gaze, and Patricia Hill Collins, whose monumental work on the outsider within contributed to how I conceptualized the coded gaze and evocative audits. My favorite part was walking through case studies of the real-world impact of research like "Gender Shades" and art like "AI, Ain't I A Woman?" that had resonated with different communities. I dropped poetic phrases as I spoke and enjoyed the pregnant pauses to wet my tongue. "Teaching truths hidden in data, each entry and omission, a person worthy of respect."

I compared the ways in which evocative audits and algorithmic audits complemented each other while enumerating their respective strengths, limitations, and risks. Evocative audits used the specificity of an individual experience to humanize algorithmic harms that reflect systemic discrimination. They incorporated counter-demo to demolish assumptions and demonstrate failures. Algorithmic audits used the generalizability of many ex-

amples to show systematic discrimination. Their strengths also
had shadow sides. Not everyone would understand the cultural
references used in an evocative audit, though those references
might resonate deeply with individuals from communities likely
to experience the harm on display.

I explained that algorithmic audits, with their detached performance metrics, were not particularly suited for showing the
human impact of algorithmic systems. Both types of audit, particularly when directed at powerful tech companies, risked retali-
ation. I warned wide-eyed undergraduates not to go up against
tech companies without having legal support, and I spoke about
the attacks I had faced. I concluded that despite the risks the re-
wards were worth it. Major tech companies, including IBM and
Amazon, had stopped or paused selling facial recognition technologies to law enforcement, and others, like Google and Microsoft, had stopped making publicly available products that attempt
to guess attributes like gender and age.

Then I turned the floor over to Ethan.

Never one to let me have an easy time, Ethan explained to the
room that there were still decisions to be made. I would need to
sufficiently answer questions from the committee and the audience. Then the committee would step outside and deliberate my
verdict. In the corner of my eye, I saw my dad glance at his watch;
he was ready to celebrate. Dr. Sweeney began the battery of
questions, followed by Hal, Catherine, and finally Ethan. As they
went through their questions I pulled up slides I had prepared in
anticipation of what they might ask to address their probes one
by one. Chance favors the prepared. Then came the audience
questions. I had tried to keep the event as small as possible but
word had gotten out. An MIT freshman had somehow heard that
my defense was taking place and went through all the pandemic

protocols to make it in person. The expression on his face reminded me of mine when I had watched Media Lab doctoral students like Xiao Xiao and Rebecca Kleinberger defend their PhDs.

These Media Labbers had set an extraordinarily high bar for the rest of us to follow, and I counted myself lucky should I be able to match their depth and breadth of work. Now I knew that the secret behind their towering dissertations was years of reading, writing, failures, experimentation, and perseverance, made possible by a circle of love. The freshman asked, "What can an undergraduate like me studying computer science do to support the work of algorithmic justice?" I hoped to one day have a book to suggest some possibilities. In that moment I responded, "Question the status quo, reach out and listen to people who experience algorithmic harms, check your assumptions, and explore intellectual terrain beyond computer science. It was learning more about subjects like intersectionality that gave me insights on how to pursue this work." His curiosity, like Agent Deb's when she reached out to me over social media to learn more about algorithmic bias, gave me hope that the next wave of computer science graduates would be better equipped to create future technologies more thoughtfully.

Then the committee members left the room to deliberate, and I fielded more questions as we waited.

"What was it like to . . ." Less than ten minutes later my committee returned.

Ethan cleared his throat and told us all: "I am happy to announce . . . Dr. Buolamwini has passed her defense."

HEARING ETHAN ADDRESS ME AS "DOCTOR" CEMENTED THE BEginning of a new season. He then told me the committee agreed that if I made my dissertation draft conform to the presentation I just gave, I would be good to go. The final stage would be revisions, but for now it was time to celebrate.

Following a ritual Ethan had started when he graduated his first PhD student, my academic older brother Dr. Nathan Matias, he had brought the plastic Civic sword he had each of his PhD students sign after a successful defense. This I expected. Next, he presented me with a new sword, a Wonder Woman sword signed by all my committee members. Holding the Civic sword, Ethan tapped each of my shoulders with it. I then crossed my Wonder Woman sword with the Civic sword. Ethan gave me a huge bear hug. I then took photos with the committee members and friends before heading to a reception prepared in the Lifelong Kindergarten group's lab space. "DR JOY" balloons cast a golden glow on a celebration cake. Makinde did the honor of cutting a first slice for me. Just a few feet away was the room that had served as the first gathering place of Yoda, Benjamin Franklin, the Big Friendly Giant, and me. Now Mitch came over and congratulated me, standing next to Natalie Rusk, who had urged me to go to the *Weapons of Math Destruction* reading by Cathy O'Neil many years before. Passion, peers, projects, and play were truly a winning combination.

I LEFT WITH A NEW SWORD AND A NEW TITLE: DR. BUOLAMWINI. To commemorate the triumph, I wrote the following poem:

TERMINAL RESISTANCE

Withstand the praise and the prison of expectations. Withdraw the temptation to mold yourself into who they thought you should be. Remember the doctor who told your worried mother, a PhD for this child is out of reach. Remember not because they were wrong, but because their eyes were too small to imagine the frail and fading body in their care contained a formidable spirit. Remember their miscalculation is one we can all make when we fail to look beyond present conditions and bleak predictions.

Withstand the pressure and the prism of demands spread across innocuous requests and insistent pleas. Withdraw your participation when you must fold your dignity, lower your stature, or diminish your worth to appease would be kingmakers. Remember the dear ones who took the calls and took the time to remind you of who you are without the crown and why they believe in you. Take the time to listen to the frail fading hearts who have forgotten the strength of their spirits.

Remember the significant small moments of acknowledgment and support. I remember the custodians who opened doors early so I could study longer. I remember the staff who gave me extra portions to show their pride and to encourage me to take bold strides. I remember the coaches who told me the truth lest they cheat me with cheap congratulations on less than my best. I remember the teachers who gave me more than was required, so I could know the elasticity of my capacities. Do not be afraid to reach higher than others dared, to stay longer when seemingly no one else cared, to fertilize the soil of toil before the vision germinates, before the sprout pushes through the terminal resistance that unleashes your power.

CHAPTER 23
CUPS OF HOPE

Eagle talons gripped a bundle of arrows and an olive branch on miniature Great Seals of the United States adorning a collection of cupcakes.* The cupcakes stood guard outside a room buzzing with excitement about the announcement from the White House. Fabian Rogers stood behind a podium as he shared his story as a Brooklyn tenant resisting the installation of facial recognition in his apartment building. He addressed an audience assembled on blue-backed chairs for a long-awaited launch. After extensive consultation, the White House was ready to release a Blueprint for an AI Bill of Rights in the fall of 2022. Out of many inputs there was finally one document. *E Pluribus Unum*.

---

* The Great Seal is the symbol of the United States. It appears on the seal of the president of the United States, which is affixed to the front of a podium when the president delivers remarks. I did not expect to see the symbol on cupcakes. It was a delicious surprise.

Before I warmed my seat, my footsteps echoed through a long corridor in the colossal Eisenhower Executive Office Building, next to the White House. The halls reminded me of the Rayburn Building, where congressional hearings took place. As a student, I raised my right hand twice before a committee to testify about the societal impacts of artificial intelligence. Each time I left I wondered if those hearings made a difference. I showed up without knowing the impact, because I felt a duty to share what my research had uncovered. I felt a duty to the Brooklyn tenants that I had promised to help. I had made a commitment to fight to prevent AI harms, to fight alongside the excoded. Today, I bore witness to the impact of years of advocacy from so many organizations and individuals willing to speak up about the perils of AI in a world enamored by the promises. One of those individuals, Tawana Petty, sat next to me with her cane resting on her right leg, purple-tinted glasses perched on her nose.

While my childhood enchantment with robots and my academic papers had brought me into the fight for algorithmic justice as an AI researcher, Tawana had a different path, with an often missing yet necessary perspective. She started thinking about the impact of technology on her community as a mother and concerned Detroit resident. Her curiosity and concern about initiatives like Project Greenlight, a citywide surveillance endeavor, fueled her advocacy. We connected over our shared love of poetry and our growing alarm about harmful use of AI systems. In 2022 she officially joined the Algorithmic Justice League. Before becoming our senior advisor on policy and advocacy, she knew what it was like to worry about making ends meet, what it meant to be without a home, and what it meant to be an organizer. She would often say to me, "I am highly educated, not highly schooled."* I love that phrase because it reminds me that credentials and degrees have their place, but they are not requirements to learn about the impacts of technology or push for change. You don't need a PhD from MIT to make a difference in the fight for algorithmic justice. All you need is a curious mind and a human heart.

You don't have to know precisely how biometric technologies work to know that when they are used for mass surveillance and invade your privacy, they do not make us safer by default. You don't have to know what a neural net is to know that if an AI system denies you a job because of your race, gender, age, disability, or skin color, something is wrong. You don't have to be an AI researcher to know that if companies take your creative work and use it to create products without permission and compensation, you have been wronged.

What we need to know is that our creative rights will be protected and that no company claiming to be responsible or ethical can sell products built with unconsented data. What we need to know is that our biometric information, like our face data and the unique sound of our voices, will be protected. Multiple cities across the United States have placed restrictions on the use of facial recognition technology by law enforcement, which has repeatedly put innocent people and Black men in particular in jail. We need federal biometric protections in the United States and across the world. After a public outcry when the Internal Revenue Service adopted an outside company to verify the faces of taxpayers, the agency made a statement that they would move

---

* When I asked her about the phrase, she said it was inspired by her mentor, civil rights leader Grace Lee Boggs, who emphasized to her that there is a difference between education and schooling.

away from using third parties for access to basic government services. We must hold them accountable to that commitment so that nobody has to give up their face to get their benefits or access veteran services. We should not have to submit invaluable data to third-party companies that require us to waive our right to pursue legal action even if we have no real choice but to use their products. We do not have to accept that if AI tools have been adopted we cannot reverse course. We do not have to accept that if companies have already created a product it is a foregone conclusion that the product will be used. In Italy, regulators put a pause on ChatGPT due to privacy concerns after a data breach. They fined Clearview AI, a company that scraped billions of face photos without consent, and they mandated the faces of Italian residents be removed from their systems. We can go further and demand that all companies creating AI systems based on personally identifiable information must prove consent has been obtained and must delete any ill-gotten data and the AI models created with unconsented data.

We can demand face purges and deep data deletions. Meta deleted more than a billion faceprints and agreed to a $650 million settlement in a legal dispute over their use of the face data uploaded by Facebook users. This action was made possible because of the Illinois Biometric Information Privacy Act, which makes it illegal to use biometric information in the state without obtaining consent. Litigation and public pushback make a difference, and so too does legislation. We need laws. Over the years draft legislation on algorithmic accountability, remote biometric technologies, and data privacy has been introduced. With growing awareness about the impact of AI on our lives, we need to know that our government institutions will protect our civil rights regardless of how technology evolves.

The AI Bill of Rights was assembled to provide an affirmative vision for the kinds of protections needed to preserve civil rights as the creation and adoption of AI systems increase. AI systems should work safely and effectively, data privacy must be protected, and automated systems should not propagate unlawful discrimination. These commonsense protections need to be both asserted and implemented. Released as a blueprint and playbook to give concrete examples for implementation, the AI Bill of Rights represents a stepping stone toward sorely needed legislation—the kind of legislation that would lead to systemic change, so we would not have to rely on the voluntary good behavior of companies.

As I welcomed the announcement of the AI Bill of Rights, I looked around the room. Seated in front of me and across the aisle were representatives from the National Institute of Standards and Technology. I had visited the NIST website often to support my research efforts, and it felt like meeting characters in a well-worn book. In January 2023, NIST would release an AI Risk Management Framework that did the important work of spelling out steps companies could take to prevent AI harms and AI discrimination in their products. These complementary efforts gave me hope. The first step to addressing a problem is acknowledging it exists. I can remember when people used to question if algorithmic bias was even real or if machines could discriminate in harmful ways, since they were supposed to be "objective."

The launch event for the AI Bill of Rights featured a panel discussion with federal officials from the Department of Education, the Department of Health and Human Services, the Equal Employment Opportunity Commission, and the Consumer Financial Protection Bureau. The panelists spoke as if the risks of AI were a forgone conclusion: "We know AI can perpetuate bias." I also remember feeling a special connection when Dr. Alondra Nelson, the acting director of the White House Office of Science and Technology Policy, took the podium and shared the motivation behind this collective effort. When we embraced, our temples connected, resting for a moment. Her hug felt like an affirmation and a thank-you for the perseverance in raising awareness about AI harms. I hope she felt my admiration for her leadership.

Hope was on the rise in Europe too. The EU AI Act was under deliberation. When passed, it would set a precedent for how AI would be governed not just in the European Union but in other parts of the world. Algorithmic risk assessments and AI audits like the one I did for the "Gender Shades" project would not just be nice-to-have ideas but requirements when AI systems were used in high-risk contexts like law enforcement, employment, and education. As part of the African diaspora, I cannot forget that AI harms are being felt in the Global South, and all too often the people experiencing the burdens are those least represented in deciding the local and international laws that govern their use. So many of the conversations and deliberations I have been a part of on how to prevent AI harms have centered the interests of the Global North. The Distributed Artificial Intelligence Research (DAIR) Institute has been especially vocal about the need to distribute resources and research on AI beyond the hands of a few large tech companies. In May 2023, more than 150 workers who provide content moderation and data detoxification services assembled in Nairobi, Kenya. They voted to establish the first African Content Moderators Union. Though often left out of global conversations, many of the Kenyan workers are paid wages of less than $2.00 an hour to go through trauma-inducing content for products like ChatGPT, TikTok, and Facebook. Their initiative to unionize and bring attention to the lack of mental health support, low pay, and unstable work is an important step in combating exploitative practices that power headline-grabbing AI products.[1]

The work toward algorithmic justice must not be just international; it must also be intergenerational. The next generation is making strides with youth-led organizations like Encode Justice, which is focused on building a movement for human-centered AI with members across thirty countries in both high school and college. We partnered with Encode Justice as the AJL began to build a harms-reporting platform for everyday people to share their experiences and seek help. We started by focusing on the use of AI systems in schools for everything from automated grading to teacher assessments and exam proctoring. We expanded to collect reports from taxpayers struggling to access essential government services, travelers feeling forced to submit to face scans, and more.[2] Beyond our efforts we need governments all over the world to maintain AI incident–reporting platforms that document known issues with AI systems to enable future learning and prevention. We must also put in mechanisms for redress so that if someone is harmed by an AI tool, they can reach out to relevant government agencies for support. Through my work, I connect with individuals and organizations committed to preventing AI harms. My work brings me into contact with companies working to build AI systems that are ethical and actively seeking ways to improve. There is a growing ecosystem of tech justice organizations, equity-minded research centers, and education initiatives that are building power, collecting evidence, and raising consciousness about the dangers that lie on the other side of AI promises.

With organizations like AJL and DAIR, led by people with

lived experience of AI harms and committed to fighting alongside the excoded, the work continues. But we cannot do it without you.

If you have a face, you have a place in the conversation and the decisions that impact your daily life—decisions that are increasingly being shaped by advancing technology that sits under the umbrella of artificial intelligence. We need your voice, because ultimately the choice about the kind of world we live in is up to us. We do not have to accept conditions and traditions that undermine our ability to have dignified lives. We do not have to sit idly by and watch the strides gained in liberation movements for racial equality, gender equality, workers' rights, disability rights, immigrant rights, and so many others be undermined by the hasty adoption of artificial intelligence that promises efficiency but further automates inequality.

THE RISING FRONTIER IN THE FIGHT FOR CIVIL RIGHTS AND human rights will require algorithmic justice, which for me ultimately means

that people have a voice and a choice in determining and shaping the algorithmic decisions that shape their lives,

that when harms are perpetuated by AI systems there is accountability in the form of redress to correct the harms inflicted,

that we do not settle on notions of fairness that do not take historical and social factors into account,

that the creators of AI reflect their societies,

that data does not destine you to unjust discrimination,

that you are not judged by the content of data profiles you never see,

that we value people over metrics,

that your hue is not a cue to dismiss your humanity,

that AI is for the people and by the people, not just the privileged few.

After the launch event, Tawana and I grabbed some cupcakes on our way out of the Executive Office Building. We paused at the top of the Navy Steps, which overlook the White House. We stood for a moment, tiny figures on silver stone, still willing to believe our tomorrows will be better than our yesterdays—this belief inspired not by machines and the progress of technology, but by the perseverance and the creativity of everyday people. The future of AI remains open-ended. Will we let power in the hands of a few tech companies dictate our lives? Will we strive for a society that protects the rights of all people? Will we dare to believe in our individual and collective power? Will we follow the drumbeat of justice? The answers are ultimately up to us.

WE FINISHED OUR CUPCAKES, REFUELED TO CONTINUE THE fight for algorithmic justice.

SEAT AT THE TABLE

A chain-link leash clung to a German shepherd with taut muscles awaiting a command. A black SUV pulled into the white security tent, and a woman with curly blond hair and a bald man with a thick graying beard stepped out. Several police officers approached the vehicle searching for explosives. Finally the car was cleared and drove up next to me. I was on the corner of Mason and California streets. Someone important would soon be at the Fairmont Hotel, and I had to move quickly. The passenger window of the SUV opened and an outstretched hand gave me an envelope that had passed the scrutiny of the guards. I took it quickly, yelled out a thank-you, and rushed to the hotel. Before I could return to the sun-soaked view of Alcatraz from my room, I had to go through another security checkpoint. Someone very important was in town. It was Monday, June 19, 2023—Juneteenth—and I had planned on being at a barbecue until I had received an email from a White House staffer the previous Wednesday. The White House was organizing a roundtable on AI, and I had been selected as one of eight experts to attend. In the lobby, guests were being pacified with wine when they learned they would not be able to return to their rooms until the Secret Service had finished their security sweep. Men with press passes walked through the lobby, some with cameras around their necks. I went to the front desk to ask about my dry cleaning. It still had not arrived and I needed my red blazer for tomorrow.

MY CLOCK READ 3:00 A.M. AS I BEAT THE SUNRISE TO IRON MY clothes and gather my talking points. A few hours later, after donning gold-framed glasses and stud earrings, I retrieved the envelope and started my search for the Gold Room on the lobby level. Seeing my uncertain footsteps, a bellman guided me to my destination, where I dutifully swabbed my nostrils to test for COVID-19. Then I was escorted to the Venetian Room, housing a historic stage where Tony Bennett once crooned "I left my heart in San Francisco." Nameplates announced the seating arrangement for the roundtable. My name was next to Governor Newsom's.

A man walked up beside me, encasing my hand in a firm and welcoming grip. "Hello, I'm Joe Biden." Governor Newsom chuckled. "We know who you are." I could not compute a response. I was in a surreal moment as I watched him greet all who were gathered. Several video cameras took in the action for a live broadcast. President Biden delivered opening remarks.

Then the press rushed in like a hungry hunting pack to take photos and shout questions, most of which related to Hunter

Biden rather than AI. The president's only response was "I'm proud of my son." The press was ushered out so we could begin the closed session.

Arati Prabhakar, the director of the White House Office of Science and Technology Policy, set the stage and introduced an expert who talked about the possibilities of AI in education: "Imagine a tutor for each student." He painted a world where students could talk to book characters simulated by chatbots, and instruction could be tailored to the interests of each child. The vision he offered was supported by promising results from an on-going pilot program. With that, Arati turned toward President Biden. "We've started with the possibilities. Now let's talk about the dangers." She introduced me and looked on with anticipation. President Biden's inquisitive gaze caught mine. I pulled out the contents of the envelope my friends had delivered the day before, and passed its contents to Governor Newsom to give to the president.

President Biden held in his hand a photo of a man with two young girls, one on his lap and the other near his knee. "Mr. President, this photo is a picture of a man named Robert Williams with his two daughters, Julia and Rose. Earlier this year I met Robert and his wife, Melissa, at the Gender Shades Justice Award. His children are looking at the Gender Shades Justice placard. This is the first award given to an individual negatively impacted by AI and fighting back. You see, Robert was arrested in front of his wife and children after a false facial recognition match. He was held in a detention center for thirty hours before seeing a police officer. The officer told him the computer had brought him up as a criminal suspect. When Robert saw the photo of the man he supposedly resembled, he said, 'It looks nothing like me. You don't think all Black people look alike?'"

President Biden interjected. "Was he arrested because he was African American?"

"Yes, Mr. President. Numerous tests of facial recognition systems have shown ongoing bias with people with darker skin." I paused. "In addition, we have documented evidence of gender bias and age bias. U.S. government tests show that some facial recognition algorithms can fail ten to one hundred times more on Asian and African American faces than white faces. And on some tests, these systems had the worst performance on . . ." My mind went blank for a second as I tried to remember my overall point while taking in the spectacle of a closed-door roundtable in an enormous room. There were eleven of us at the table but we were surrounded by what appeared to be twenty staffers, a cadre of hotel staff, and the president's press secretary. I recovered my thoughts. "In some cases, some of these systems perform the worst on Native Americans." Before I could dive deep into numbers on gender bias and age bias, President Biden had more questions. The most important one was: "What should I do?"

I took a deep breath. What an honor it was to have my expertise and recommendations sought after by the most powerful man in the world. A man in his eighties, forty-seven years my senior, with a white crown of hair and porcelain skin, the historic symbol of power. A man who had been vice-president to President Obama, the first Black man to hold the presidency of the United States. A man who, according to Governor Newsom, had forgotten more things about the Senate than we will ever know. I now sat two seats away in my red blazer and with my curly hair in a natural style, ready to meet the moment.

"Mr. President, the United States has an opportunity to lead on biometric rights, but right now we are falling behind. Just last week European lawmakers voted to push forward the EU AI Act, which restricts the use of live facial recognition in public spaces because of the discriminatory and invasive impacts of biometrics. The U.S. is going the opposite way, with the Transportation Security Administration [TSA] piloting a program to use facial recognition at domestic airports. According to the public TSA road map, the agency's plan is to have this pilot expand to all airports and become the default way of traveling. We must stop this escalation. And here is an issue with bipartisan support. When I testified in front of Congress in 2019, Representative Jim Jordan and AOC were in agreement on the need to push back on facial recognition."

President Biden opened his hands. "I have to leave—I can't believe this. Those two in agreement?" Murmurs of laughter rippled around the table.

I added, "The United States should make sure the face is not the final frontier of privacy. We should lead on biometric rights and stop the use of facial recognition by the TSA for domestic flights specifically, but also its use for mass surveillance more broadly."

I remember the president positioning his hand to take notes, and then Arati directed his attention to another roundtable participant. I listened intently, repositioning my body based on Governor Newsom's movements. When he leaned forward I had to lean back, as I could not see over his six-foot frame and gelled hair even when he was seated. I was especially captured by the remarks of Nobel Laureate Jennifer Doudna, the inventor of CRISPR, the scientific breakthrough that enabled precise gene editing: "As the biologist at the table, I want to turn your attention to lessons we learned as we grapple with the benefits and perils of a powerful technology." The ability to alter genes had many medical benefits. Earlier in the Gold Room, as we awaited our COVID test results, Dr. Doudna had shared with me her excitement over the successful introduction of a genomic therapy for sickle cell disease. With only one treatment they could cure the condition, which impacts around a hundred thousand individuals in the United States. The costs were still incredibly high and her nonprofit was working on approaches to bring them down and make the therapies more accessible. The technology was saving lives and had the promise to save even more. But with every great power comes a shadow side, and here she turned our attention. Because gene editing can allow us to change the genetic composition of our species, her field had to grapple with deep ethical questions, and it had to erect essential safeguards to keep its power in check. "Mr. President," she continued, "we convened the leaders in the field in a series of international summits to establish the guardrails in conversation. Consider establishing summits that set the global norms of this technology."

The hour passed quickly as other experts shared their priority areas for the president to direct federal funding, with a focus on shoring up AI talent in the United States, strengthening the public sector's AI capabilities, and supporting projects to develop assistive technologies using AI. President Biden considered these perspectives while giving us insights into the political realities of getting Congress to approve any new spending or pass laws. With thirty seconds to go, the final expert offered a narrative to gain support for investing in the ideas offered around the table. "Mr. President, you don't know much about technology—and that's a good thing, because most people can relate to that—but you do care very deeply for the future of this nation. The next generation stands to be the greatest beneficiaries of the benefits of AI or the greatest losers if we do not act." Just as quickly as the event had started, President Biden left the same way he entered, offering warm handshakes and asking us to continue to advise. "I'm like your poor relative, I will ask for your help and offer no money in return," he joked. Arati and the sleep-deprived staffers looked pleased with the meeting. "He usually doesn't stay overtime. He was very engaged," Arati whispered in my ear. After a group photo, we left the Venetian Room with police officers observing our movements.

As I stepped out of the elevator onto the twenty-second floor, Ahcene Mklat, an Algerian immigrant, escorted me. I had taken the advice of my father, who encouraged me to do my best to treat everyone with dignity. Ahcene, I'd thought, was just a friendly man at the restaurant when we chatted the night before. He turned out to be a manager at the Fairmont, and when he heard I needed to change rooms because of a barking dog in the neighboring room, he brought me to one of the best, the Diplomat Suite, also known as the Tony Bennett Suite, room 2211. The penthouse suite was occupied by someone who could not be named. In my room, I thought about what Governor Newsom had said, that here in San Francisco so many of the companies fueling the AI evolution are within forty-seven miles. I slipped off my red blazer and sunk into the bed. I looked at the sweeping panoramic view from the multi-room suite, seeing my reflection overlaid on a glittering skyline. My mind was occupied by hope. Sleep cradled me into California dreaming.

UNSTABLE DESIRE

Prompted to competition
Where be the guardrails now?
Threat in sight
Will might make right?

Hallucinations
Taken as prophecy

Destabilized
On a middling journey
To outpace
To open chase
To claim supremacy
To reign indefinitely

Haste and Paste
Control Altering Deletion
Unstable Desire
Remains Undefeated
The Fate of AI
Still Uncompleted

Responding with fear
Responsible AI, beware
Profits do snare
People still dare
To believe our humanity
Is more than neural nets
And transformations of
Collected muses
More than data and errata
More than transactional diffusions
Are we not transcendent beings
Bound in transient forms?

Can this power
be guided with care?
Augmenting delight alongside

Economic destitution?
Temporary Band-Aids cannot
Hold the wind when the task
Ahead is to transform the
Atmosphere of innovation.

The android dreams entice
The nightmare schemes of vice.

—POET OF CODE, CERTIFIED HUMAN-MADE

ACKNOWLEDGMENTS

Thank you to the ineffable source of life and creation that makes all things possible. Thank you to the ancestors who came before me and set foundations that make my journey as a poet of code possible.

Thank you to my Queen Mother, Frema, and Daring Father, the first Dr. J. Buolamwini, who taught me the beauty of searching for truth through both art and science and the importance of standing by my convictions. Thank you to my Nana Joyce, who stemmed my first name and my first language, teaching me the greatest legacy is the imprints of our love. Thank you to my grandfather Dr. Dwuma-Badu, who passed before me and set the standard. I am proud to continue his academic legacy into a third generation. Thank you to my beloved brother, Coach B, whom I would get in trouble with for not mentioning.

Thank you to my circle of loving friends who have supported me in the sunshine and rain and bore witness to the book writing process. Thank you especially to Alicia Chong-Rodriguez, who encouraged me to persevere each time I contemplated the MIT dropout story; Dr. Naseemah Mohamed for more than words can describe; Dr. Aubry Threlkeld, Dr. Kevin Hu, and Olumakinde Ogunnaike for providing listening ears as I worked to articulate the boundaries of many of the ideas contained in this book; and Dr. Sandy Henin and Shelby Clark for the laughter and cherished memories clearing many heights.

Thank you to my sister Face Queens Dr. Timnit Gebru and Agent Deb Raji, whose intellectual companionship and commitment to algorithmic justice have made my work possible. Thank you to my PhD sisters, Dr. Alexis Hope and Jaleesa Trapp, who made earning our degrees together at the MIT Media Lab a fun adventure of magic and mischief.

Thank you to the Center for Civic Media (rest in power) for showing me what it means to do engaged scholarship. Thank you to the MIT Media Lab for the freedom to create on the edges.

Thank you to the teams at P&G Olay and Badger Agency for elevating and celebrating my work as an artist, academic, and young Black woman leader to inspire more girls and women to pursue STEM.

Thank you to the Algorithmic Justice League team, funders, and advisory board for making it possible for me to write a book while building an organization.

Honorary guardians of the Algorithmic Justice League Professor Ethan Zuckerman, Professor Hal Abelson, Professor Latanya Sweeney, and Professor Catherine D'Ignazio are true heroes. Thank you for encouraging me to prioritize my health, for your generosity with your time, and for the momentous Dr. Justice sword of knowledge.

Finally, this book would not be possible without the incredible members of my book team, including my literary agent, Mollie Glick; my book coach, Alexis Gargagliano; and my editor, Marie Pantojan, who guided me over an unforgettable three-year journey.

NOTES

INTRODUCTION

1. Guia Marie Del Prado, "Stephen Hawking, Elon Musk, Steve Wozniak and Over 1,000 AI Researchers Cosigned an Open Letter to Ban Killer Robots," *Business Insider,* July 28, 2015, www.businessinsider.in/latest/stephen-hawking-elon-musk-steve-wozniak-and-over-1000-ai-researchers-co-signed-an-open-letter-to-ban-killer-robots/articleshow/48246087.cms.

2. Krystal Hu, "ChatGPT Sets Record for Fastest-Growing User Base—Analyst Note," Reuters, February 2, 2023, www.reuters.com/technology/chatgpt-sets-record-fastest-growing-user-base-analyst-note-2023-02-01/.

3. Dina Bass, "Microsoft Invests $10 Billion in ChatGPT Maker OpenAI," Bloomberg, January 23, 2023, https://www.bloomberg.com/news/articles/2023-01-23/microsoft-makes-multibillion-dollar-investment-in-openai.

4. Jennifer Korn, "Getty Images Suing the Makers of Popular AI Art

Tool for Allegedly Stealing Photos," CNN, January 18, 2023, www
.cnn.com/2023/01/17/tech/getty-images-stability-ai-lawsuit/index
.html.

5. Simon Ellery, "Fake Photos of Pope Francis in a Puffer Jacket Go
Viral, Highlighting the Power and Peril of AI," CBS News, March
28, 2023, www.cbsnews.com/news/pope-francis-puffer-jacket-fake
-photos-deepfake-power-peril-of-ai/.

6. Vanessa Yurkevich, "Universal Music Group Calls AI Music a
'Fraud,' Wants It Banned from Streaming Platforms. Experts Say
It's Not That Easy," CNN, April 18, 2023, www.cnn.com/2023/
04/18/tech/universal-music-group-artificial-intelligence/index
.html.

7. "Mom Warns of Hoax Using AI to Clone Daughter's Voice," ABC
News, April 13, 2023, abcnews.go.com/GMA/Family/mom-warns-
hoax-ai-clone-daughters-voice/story?id=98551351.

8. Sidney Fussell, "A Flawed Facial-Recognition System Sent This Man
to Jail," Wired, June 24, 2020, www.wired.com/story/flawed-facial
-recognition-system-sent-man-jail/.

9. "Dutch Student Files Complaint with the Netherlands Institute
for Human Rights About the Use of Racist Software by Her
University," Racism and Technology Center, July 28, 2022,
racismandtechnology.center/2022/07/28/dutch-student-files
-complaint-with-the-netherlands-institute-for-human-rights-about
-the-use-of-racist-software-by-her-university/.

10. Melissa del Bosque, "Facial Recognition Bias Frustrates Black Asylum Applicants to US, Advocates Say," The Guardian, February 8,
2023, www.theguardian.com/us-news/2023/feb/08/us
-immigration-cbp-one-app-facial-recognition-bias.

11. Lauren Walker, "Belgian Man Dies by Suicide Following
Exchanges with Chatbot," The Brussels Times, March 28, 2023,
www.brusselstimes.com/430098/belgian-man-commits-suicide
-following-exchanges-with-chatgpt.

PART 1

CHAPTER 3

1. Clare Garvie, Alvaro Bedoya, and Jonathan Frankle, "The Perpetual Line-up: Unregulated Police Face Recognition in America," Georgetown Law Center on Privacy and Technology, October 18, 2016, www.perpetuallineup.org/.

PART II

CHAPTER 5

1. Ziad Obermeyer, Brian Powers, Christine Vogeli, and Senhil Mullainathan, "Dissecting Racial Bias in an Algorithm Used to Manage the Health of Populations," *Science* 366, no. 6464 (October 25, 2019): 447–53, www.science.org/doi/10.1126/science.aax2342.

2. Crina Grosan and Ajith Abraham, "Rule-Based Expert Systems," in *Intelligent Systems,* Intelligent Systems Reference Library, vol. 17 (Berlin: Springer, 2011), 149–85, link.springer.com/chapter/10.1007/978-3-642-21004-4_7.

3. "701 Translator," IBM press release, January 8, 1954, www.ibm.com/ibm/history/exhibits/701/701_translator.html.

4. Thierry Poibeau, "The 1966 ALPAC Report and Its Consequences" [1966], in *Machine Translation* (Cambridge, MA: MIT Press, 2017), 75–89, ieeexplore.ieee.org/document/8093945.

5. John McCarthy, Marvin L. Minsky, Nathaniel Rochester, and Claude E. Shannon, "A Proposal for the Dartmouth Summer Research Project on Artificial Intelligence, August 31, 1955," *AI Magazine* 27, no. 4 (Winter 2006): 12, ojs.aaai.org/index.php/aimagazine/article/view/1904.

CHAPTER 6

1. Keyon Vafa, Christian Haigh, Alvin Leung, and Noah Yonack, "Price Discrimination in the Princeton Review's Online SAT Tutoring Service," *Technology Science,* August 31, 2015, techscience.org/a/2015090102/.

2. Julia Angwin, Surya Mattu, and Jeff Larson, "The Tiger Mom Tax: Asians Are Nearly Twice as Likely to Get a Higher Price from Princeton Review," ProPublica, September 1, 2015, www .propublica.org/article/asians-nearly-twice-as-likely-to-get-higher -price-from-princeton-review.

3. Joanna Stern, "Department Store Mannequins Are Watching You. No, Really," ABC News, November 26, 2012, abcnews.go.com/ Technology/department-store-mannequins-watch-eyesee-analyzes -shoppers-webcams/story?id=17813441.

4. "Facing Facts: Best Practices for Common Uses of Facial Recognition Technologies," Federal Trade Commission, October 2012, www.ftc.gov/reports/facing-facts-best-practices-common-uses -facial-recognition-technologies.

5. Alex Hern, "Anti-Surveillance Clothing Aims to Hide Wearers from Facial Recognition," *The Guardian*, January 4, 2017, www .theguardian.com/technology/2017/jan/04/anti-surveillance -clothing-facial-recognition-hyperface.

6. Inioluwa Deborah Raji, I. Elizabeth Kumar, Aaron Horowitz, and Andrew Selbst, "The Fallacy of AI Functionality," in *2022 ACM Conference on Fairness, Accountability, and Transparency* (FAccT '22), June 21–24, 2022, Seoul, Republic of Korea (New York: ACM, 2022), 959–72, dl.acm.org/doi/fullHtml/10.1145/3531146.3533158.

7. Scott Pelley, "Is Artificial Intelligence Advancing Too Quickly? What AI Leaders at Google Say," *60 Minutes*, CBS News, April 16, 2023, www.cbsnews.com/news/google-artificial-intelligence -future-60-minutes-transcript-2023-04-16.

8. Erik Learned-Miller, Vicente Ordóñez, Jamie Morgenstern, and Joy Buolamwini, "Facial Recognition Technologies in the Wild: A Call for a Federal Office," Algorithmic Justice League, May 29, 2020, www.ajl.org/federal-office-call.

9. Xiaolin Wu and Xi Zhang, "Automated Inference on Criminality Using Face Images," arXiv, May 26, 2017, arxiv.org/abs/1611.04135v1.

10. "Advances in AI Are Used to Spot Signs of Sexuality," *The Economist*, September 9, 2017, www.economist.com/science-and-technology/ 2017/09/09/advances-in-ai-are-used-to-spot-signs-of-sexuality.

11.   Lucas Ramón Mendos, Kellyn Botha, Rafael Carrano Lelis, Enrique López de la Peña, Ilia Savelev, and Daron Tan, *State-Sponsored Homophobia 2020: Global Legislation Overview Update* (Geneva: ILGA, December 2020), ilga.org/downloads/ILGA_World_State_Sponsored_Homophobia_report_global_legislation_overview_update_December_2020.pdf.

12.   Jeffrey Dastin, "Rite Aid Deployed Facial Recognition Systems in Hundreds of U.S. Stores," Reuters, July 28, 2020, www.reuters.com/investigates/special-report/usa-riteaid-software/.

13.   "Face Off," Big Brother Watch, May 2018, bigbrotherwatch.org.uk/campaigns/stop-facial-recognition/report/.

14.   "Passport Facial Recognition Checks Fail to Work with Dark Skin," BBC News, October 9, 2019, www.bbc.com/news/technology-49993647.

15.   Morgan Meaker, "This Student Is Taking On 'Biased' Exam Software," *Wired*, April 5, 2023, www.wired.com/story/student-exam-software-bias-proctorio/.

CHAPTER 8

1.   Gary B. Huang, Manu Ramesh, Marwin Mattar, and Erik Learned-Miller, "Labeled Faces in the Wild: A Database for Studying Face Recognition in Unconstrained Environments," Workshop on Faces in "Real-Life" Images: Detection, Alignment, and Recognition, October 2008, people.cs.umass.edu/~elm/papers/lfw.pdf.

2.   Patrick Grother and Mei Ngan, "Face Recognition Vendor Test (FRVT): Performance of Face Identification Algorithms," NIST Interagency Report 8009, May 26, 2014, nvlpubs.nist.gov/nistpubs/ir/2014/NIST.IR.8009.pdf.

3.   Hu Han and Anil K. Jain, "Age, Gender and Race Estimation from Unconstrained Face Images," MSU Technical Report, 2014, biometrics.cse.msu.edu/Publications/Face/HanJain_UnconstrainedAgeGenderRaceEstimation_MSUTechReport2014.pdf.

4.   Rasmus Rothe, Radu Timofte, and Luc Van Gool, "Deep Expectation of Real and Apparent Age from a Single Image Without Facial

Landmarks," *International Journal of Computer Vision* 126, nos. 2–4 (2018): 144–57, data.vision.ee.ethz.ch/cvl/rrothe/imdb-wiki/.

5. Kimberlé Crenshaw, "Demarginalizing the Intersection of Race and Sex: A Black Feminist Critique of Antidiscrimination Doctrine, Feminist Theory and Antiracist Politics," *University of Chicago Legal Forum* 1989, no. 1 (1989): Article 8, chicagounbound.uchicago.edu/uclf/vol1989/iss1/8.

6. Joy Buolamwini, "Gender Shades: Intersectional Phenotypic and Demographic Evaluation of Face Datasets and Gender Classifiers," Master's thesis, MIT, 2017, 61, dspace.mit.edu/handle/1721.1/114068.

7. Jeffrey Dastin, "Amazon Scraps Secret AI Recruiting Tool That Showed Bias Against Women," Reuters, October 10, 2018, www.reuters.com/article/us-amazon-com-jobs-automation-insight-idUSKCN1MK08G.

8. "Women in Politics: 2017," UN Women report, January 1, 2017, www.unwomen.org/sites/default/files/Headquarters/Attachments/Sections/Library/Publications/2017/FemmesEnPolitique_2017_English_Web.pdf.

9. Brendan F. Klare, Ben Klein, Emma Taborsky, Austin Blanton, Jordan Cheney, Kristen Allen, Patrick Grother, Alan Mah, and Anil K. Jain, "Pushing the Frontiers of Unconstrained Face Detection and Recognition: IARPA Janus Benchmark A," *2015 IEEE Conference on Computer Vision and Pattern Recognition (CVPR)* (Boston, MA: IEEE, 2015), 1931–1939, www.cv-foundation.org/openaccess/content_cvpr_2015/papers/Klare_Pushing_the_Frontiers_2015_CVPR_paper.pdf.

10. "Mid-year Population Estimates 2017," Statistics of South Africa, July 31, 2017, www.statssa.gov.za/publications/P0302/P03022017.pdf.

PART III

CHAPTER 9

1. "What Census Calls Us: A Historical Timeline," Pew Research Center, www.pewresearch.org/wp-content/uploads/2020/02/PH
_15.06.11_MultiRacial-Timeline.pdf.
2. "Australian Standard Classification of Cultural and Ethnic Groups (ASCCEG)," Australian Bureau of Statistics, December 18, 2019, www.abs.gov.au/statistics/classifications/australian-standard-classification-cultural-and-ethnic-groups-ascceg/latest-release.
3. Vishad Gupta and Vinod Kumar Sharma, "Skin Typing: Fitzpatrick Grading and Others," *Clinics in Dermatology* 37, no. 5 (September 1, 2019): 430–36, doi.org/10.1016/j.clindermatol.2019.07.010; Rita Oliveira, Joana Ferreira, Luis Felipe Azevedo, and Isabel F. Almeida, "An Overview of Methods to Characterize Skin Type: Focus on Visual Rating Scales and Self-Report Instruments," *Cosmetics* 10, no. 1 (January 2023): 14, www.mdpi.com/2079-9284/10/1/14.
4. Arucha Treesirichod, Somboon Chansakulporn, and Pattra Wattanapan, "Correlation Between Skin Color Evaluation by Skin Color Scale Chart and Narrowband Reflectance Spectrophotometer," *Indian Journal of Dermatology* 59, no. 4 (July/August 2014): 339–42, www.ncbi.nlm.nih.gov/pmc/articles/PMC4103266/.
5. Morgan Klaus Scheuerman, Jacob M. Paul, Jed R. Brubaker, "How Computers See Gender: An Evaluation of Gender Classification in Commercial Facial Analysis Services," *Proceedings of the ACM on Human-Computer Interaction* 3, issue CSCW (November 2019): 1–33, dl.acm.org/doi/abs/10.1145/3359246.
6. These findings come from the author's 2017 MIT master's thesis, "Gender Shades," www.media.mit.edu/publications/full-gender-shades-thesis-17/.
7. "Women in National Parliaments," Inter-Parliamentary Union, March 1, 2017. In the chart, 193 countries are listed in descending order according to the percentage of women in that country's lower or single house: archive.ipu.org/wmn-e/arc/classif010317.htm.

8.   Olivia Solon, "Facial Recognition's 'Dirty Little Secret': Millions of Online Photos Scraped Without Consent," NBC News, March 12, 2019, www.nbcnews.com/tech/internet/facial-recognition-s-dirty-little-secret-millions-online-photos-scraped-n981921.

9.   Olivia Solon and Cyrus Farivar, "Millions of People Uploaded Photos to the Ever App. Then the Company Used Them to Develop Facial Recognition Tools," NBC News, May 9, 2019, www.nbcnews.com/tech/security/millions-people-uploaded-photos-ever-app-then-company-used-them-n1003371.

10.   Melissa Heikkilä, "The Viral AI Avatar App Lensa Undressed Me—Without My Consent," MIT Technology Review, December 12, 2022, www.technologyreview.com/2022/12/12/1064751/the-viral-ai-avatar-app-lensa-undressed-me-without-my-consent/.

11.   Olivia Snow, "'Magic Avatar' App Lensa Generated Nudes from My Childhood Photos," Wired, December 7, 2022, www.wired.com/story/lensa-artificial-intelligence-csem/.

12.   Romain Beaumont, "LAION-5B: A New Era of Open Large-Scale Multimodal Datasets," Laion, March 31, 2022, github.com/LAION-AI/laion5B-paper.

13.   Victoria Cavaliere, "Judge Approves $650 Million Settlement of Facebook Privacy Lawsuit Linked to Facial Photo Tagging," Business Insider, February 27, 2021, www.businessinsider.com/facebook-settlement-pay-650-million-privacy-lawsuit-biometrics-face-tagging-2021-2.

14.   "Facial Recognition: Italian SA Fines Clearview AI EUR 20 Million," European Data Protection Board, March 10, 2022, edpb.europa.eu/news/national-news/2022/facial-recognition-italian-sa-fines-clearview-ai-eur-20-million_en.

15.   Geoff Colvin, "How Amazon Grew an Awkward Side Project into AWS, a Behemoth That's Now 4 Times Bigger Than Its Original Shopping Business," Fortune, November 30, 2022, fortune.com/longform/amazon-web-services-ceo-adam-selipsky-cloud-computing/.

CHAPTER 10

1. Audrey Elisa Kerr, "The Paper Bag Principle: Of the Myth and the Motion of Colorism," *Journal of American Folklore* 118, no. 469 (Summer 2005): 271–89, www.jstor.org/stable/4137914.

2. Kenneth B. Clark and Mamie B. Clark, "Emotional Factors in Racial Identification and Preference in Negro Children," *Journal of Negro Education* 19, no. 3 (Summer 1950): 341–50, www.jstor.org/stable/2966491.

3. "Brown v. Board and 'The Doll Test,'" Legal Defense Fund, www.naacpldf.org/brown-vs-board/significance-doll-test/.

4. "Mid-Year Population Estimates 2017," Statistics of South Africa, July 31, 2017, www.statssa.gov.za/publications/P0302/P03022017.pdf.

CHAPTER 12

1. At that time it was called the FAT* Conference. The name was later changed to avoid body shaming.

2. Latanya Sweeney, "Discrimination in Online Ad Delivery," *Queue* 11, no. 3 (March 1, 2013): 10–29, dataprivacylab.org/projects/onlineads/1071-1.pdf.

3. "The Global 2000 River Blindness Program of the Carter Center," The Carter Center, December 1996, www.cartercenter.org/documents/doc237.html.

4. "River Blindness Elimination Program," The Carter Center, www.cartercenter.org/health/river_blindness/index.html.

5. Devin Coldewey, "Anthropic's Quest for Better, More Explainable AI Attracts $580M," TechCrunch, April 29, 2022, techcrunch.com/2022/04/29/anthropics-quest-for-better-more-explainable-ai-attracts-580m/.

6. Salman Ahmed, Cameron T. Nutt, Nwamaka D. Enenaya, et al., "Examining the Potential Impact of Race Multiplier Utilization in Estimated Glomerular Filtration Rate Calculation on African-American Care Outcomes," *Journal of General Internal Medicine* 36, 464–71 (2021), doi.org/10.1007/s11606-020-06280-5.

7. Emanuel Martinez and Lauren Kirchner, "Denied: The Secret Bias

Hidden in Mortgage-Approval Algorithms," *The Markup,* August 25, 2021, themarkup.org/denied/2021/08/25/the-secret-bias-hidden-in-mortgage-approval-algorithms.

8. Aaron Rieke and Miranda Bogen, "Help Wanted: An Examination of Hiring Algorithms, Equity, and Bias," Upturn, December 10, 2018, www.upturn.org/work/help-wanted/.

9. Lauren Kirchner and Matthew Goldstein, "How Automated Background Checks Freeze Out Renters," *The New York Times,* May 28, 2020, www.nytimes.com/2020/05/28/business/renters-background-checks.html.

PART IV

Portions of my PhD dissertation are incorporated into this part of the book to explain the concept of the counter-demo and provide examples of counter-demos developed by others, as well as to examine the use of the "Gender Shades" findings to support facial recognition resistance campaigns and reflect on ways my research findings can also be used to support surveillance agendas.

CHAPTER 13

1. Steve Lohr, "Facial Recognition Is Accurate, If You're a White Guy," *The New York Times,* February 9, 2018, www.nytimes.com/2018/02/09/technology/facial-recognition-race-artificial-intelligence.html.

CHAPTER 14

1. Sean Hollister, "Google Contractors Reportedly Targeted Homeless People for Pixel 4 Facial Recognition," *The Verge,* October 2, 2019, www.theverge.com/2019/10/2/20896181/google-contractor-reportedly-targeted-homeless-people-for-pixel-4-facial-recognition.

CHAPTER 15

1.   Richard Evans and Jim Gao, "DeepMind AI Reduces Google Data
     Centre Cooling Bill by 40%," Google DeepMind, July 20, 2016,
     www.deepmind.com/blog/deepmind-ai-reduces-google-data
     -centre-cooling-bill-by-40.

2.   Khari Johnson, "How Wrongful Arrests Based on AI Derailed 3
     Men's Lives," *Wired,* March 7, 2022, www.wired.com/story/
     wrongful-arrests-ai-derailed-3-mens-lives/.

3.   Kate Conger, "Amazon Workers Demand Jeff Bezos Cancel Face
     Recognition Contracts with Law Enforcement," Gizmodo, June 21,
     2018, gizmodo.com/amazon-workers-demand-jeff-bezos-cancel
     -face-recognitio-1827037509; coalition letter to Jeff Bezos, May 22,
     2018, www.aclunc.org/docs/20180522_AR_Coalition_Letter.pdf.

4.   Joy Buolamwini, "Amazon's Symptoms of FML—Failed Machine
     Learning—Echo the Gender Pay Gap and Policing Concerns," MIT
     Media Lab, October 19, 2018, medium.com/mit-media-lab/
     amazons-symptoms-of-fml-failed-machine-learning-echo-the
     -gender-pay-gap-and-policing-concerns-3de9553d9bd1.

5.   Natasha Singer, "Amazon Is Pushing Facial Technology That a
     Study Says Could Be Biased," *The New York Times,* January 24, 2019,
     www.nytimes.com/2019/01/24/technology/amazon-facial
     -technology-study.html.

6.   Matt Wood, "Thoughts on Recent Research Paper and Associated
     Article on Amazon Rekognition," *Amazon SageMaker,* January 26,
     2019, aws.amazon.com/blogs/machine-learning/thoughts-on
     -recent-research-paper-and-associated-article-on-amazon-rekognition/.

7.   Phil Stewart, "Microsoft Beats Amazon for Pentagon's $10 Billion
     Cloud Computing Contract," Reuters, October 26, 2019, www
     .reuters.com/article/us-pentagon-jedi/microsoft-beats-amazon-for
     -pentagons-10-billion-cloud-computing-contract-idUSKBN1X42IU.

8.   Yochai Benkler, "Don't Let Industry Write the Rules for AI," *Nature*
     569, no. 7755 (May 9, 2019): 161, www.nature.com/articles/d41586
     -019-01413-1.

9.   "NIST Study Evaluates Effects of Race, Age, Sex on Face Recogni-
     tion Software," NIST, December 19, 2019, www.nist.gov/news

-events/news/2019/12/nist-study-evaluates-effects-race-age-sex
-face-recognition-software.

CHAPTER 16

1. "Amicus Support Letter: Brooklyn Tenants Protest Against Facial
   Recognition Entry Systems"; the letter cites a StoneLock white
   paper that is no longer available online, docs.google.com/document/
   d/1WjcVT2kCLAEC_rS9eJcCC5S75e2VszkKkNYevWnGH3U/edit.

2. Cynthia M. Cook, John J. Howard, Yevgeniy B. Sirotin, Jerry L. Tip-
   ton, and Arun R. Vemury, "Demographic Effects in Facial Recognition and Their Dependence on Image Acquisition: An Evaluation
   of Eleven Commercial Systems," *IEEE Transactions on Biometrics, Behavior, and Identity Science* 1, no. 1 (January 2019): 32–41, ieeexplore
   .ieee.org/document/8636231; Brendan F. Klare, Mark J. Burge,
   Joshua C. Klontz, Richard W. Vorder Bruegge, and Anil K. Jain,
   "Face Recognition Performance: Role of Demographic Information," *IEEE Transactions on Information Forensics and Security* 7, no. 6
   (December 2012): 1789–1801, ieeexplore.ieee.org/document/
   6327355. Note that Brendan Klare is a former FBI facial recognition
   expert.

3. Kathleen C. Fraser, Jed A. Meltzer, and Frank Rudzicz, "Linguistic
   Features Identify Alzheimer's Disease in Narrative Speech," *Journal
   of Alzheimer's Disease* 49 (2016): 407–22, www.cs.toronto.edu/
   ~kfraser/Fraser15-JAD.pdf.

4. Clare Garvie, Alvaro Bedoya, and Jonathan Frankle, "The Perpetual
   Line-up: Unregulated Police Face Recognition in America," Georgetown Law Center on Privacy and Technology, October 18, 2016,
   www.perpetuallineup.org/.

5. Reza Shoja Ghiass, Ognjen Arandjelovic, Abdelhakim Bendada, and
   Xavier Maldague, "Infrared Face Recognition: A Comprehensive
   Review of Methodologies and Databases," *Pattern Recognition* 47,
   no. 9 (September 2014): 2807–24, arxiv.org/pdf/1401.8261.pdf.

6. "Face Surveillance and Racial Bias," ACLU Massachusetts informa-
   tional pamphlet, www.aclum.org/sites/default/files/field
   _documents/racial_bias_and_fs.pdf.

CHAPTER 17

1. Khari Johnson, "How Wrongful Arrests Based on AI Derailed 3
    Men's Lives," *Wired*, March 7, 2022, www.wired.com/story/
    wrongful-arrests-ai-derailed-3-mens-lives/.

CHAPTER 18

1. Elizabeth Kim, "Brooklyn Landlord Does an About Face on Facial
    Recognition Plan," *Gothamist,* November 21, 2019, gothamist.com/
    news/brooklyn-landlord-does-about-face-facial-recognition-plan;
    Yasmin Gagne, "How We Fought Our Landlord's Secretive Plan for
    Facial Recognition—and Won," *Fast Company,* November 22, 2019,
    www.fastcompany.com/90431686/our-landlord-wants-to-install
    -facial-recognition-in-our-homes-but-were-fighting-back.

PART V

CHAPTER 19

1. "2020 Doctorate Recipients from U.S. Universities," National Cen-
    ter for Science and Engineering Statistics, Directorate for Social, Be-
    havioral and Economic Sciences, and National Science Foundation,
    November 2021, ncses.nsf.gov/pubs/nsf22300/assets/report/
    nsf22300-report.pdf.
2. "How Is the Founding of MIT Connected to Slavery?" MIT and
    Slavery, MIT Libraries, libraries.mit.edu/mit-and-slavery/founding
    -mit/.

CHAPTER 21

1. Joy Buolamwini, "The Algorithmic Justice League," MIT Media
    Lab, December 14, 2016, medium.com/mit-media-lab/the
    -algorithmic-justice-league-3cc4131c5148.
2. Patrice Taddonio, "How an AI Scientist Turned Her Breast Cancer
    Diagnosis into a Tool to Save Lives," *Frontline,* PBS, November 4,
    2019, www.pbs.org/wgbh/frontline/article/how-an-ai-scientist
    -turned-her-breast-cancer-diagnosis-into-a-tool-to-save-lives/.

3. George Joseph and Kenneth Lipp, "IBM Used NYPD Surveillance Footage to Develop Technology That Lets Police Search by Skin Color," *The Intercept,* September 6, 2018, theintercept.com/2018/09/06/nypd-surveillance-camera-skin-tone-search/.

4. Joy Buolamwini, "When the Robot Doesn't See Dark Skin," *The New York Times,* June 21, 2018, www.nytimes.com/2018/06/21/opinion/facial-analysis-technology-bias.html.

CHAPTER 23

1. Billy Perrigo, "150 African Workers for ChatGPT, TikTok and Facebook Vote to Unionize at Landmark Nairobi Meeting," *Time,* May 1, 2023, time.com/6275995/chatgpt-facebook-african-workers-union/.

2. Joy Buolamwini, "The IRS Should Stop Using Facial Recognition," *The Atlantic,* January 27, 2022, www.theatlantic.com/ideas/archive/2022/01/irs-should-stop-using-facial-recognition/621386/; Wilfred Chan, "Exclusive: TSA to Expand Its Facial Recognition Program to Over 400 Airports," *Fast Company,* June 30, 2023, www.fastcompany.com/90918235/tsa-facial-recognition-program-privacy.

Dr. Joy Buolamwini is the founder of the Algorithmic Justice League, a groundbreaking researcher, a bestselling author, and a renowned speaker. Her writing has been featured in publications including *TIME*, *The New York Times*, *Harvard Business Review*, and *The Atlantic*. As the Poet of Code, she creates art to illuminate the impact of artificial intelligence on society and advises world leaders on preventing AI harms. She is the recipient of notable awards, including the Rhodes Scholarship, the inaugural Morals & Machines Prize, and the Technological Innovation Award from the Martin Luther King, Jr. Center for Nonviolent Social Change, and has received honorary degrees from Knox College and Dartmouth College. Her MIT research on facial recognition technologies is featured in the Emmy-nominated documentary *Coded Bias*. Born in Canada to Ghanaian immigrants, she lives in Cambridge, Massachusetts.

poetofcode.com
donate.ajl.org
X: @jovialjoy
X: @AJLUnited
Instagram: @poetofcode